WHEN THE BEACHES TREMBLED

The Incredible True Story of Stephen Ganzberger and the LCIs in World War II

ZACH S. MORRIS

When the Beaches Trembled: The Incredible True Story of Stephen Ganzberger and the LCIs in World War II

Copyright © 2022 by Zach S. Morris

Los Angeles, CA

ISBN 979-8-9851613-0-4

ISBN 979-8-9851613-1-1 (ebook)

Self-published by Zach S. Morris

First printing, 2022

Second printing, 2022

Third printing, 2025

For Grandpa

For my mother, Heidi

And for the Greatest Generation

Eternal Father, strong to save,
Whose arm hath bound the restless wave,
Who bidd'st the mighty ocean deep
Its own appointed limits keep,
O hear us when we cry to Thee
For those in peril on the sea!

Excerpts, "Eternal Father, Strong to Save"
Navy Hymn

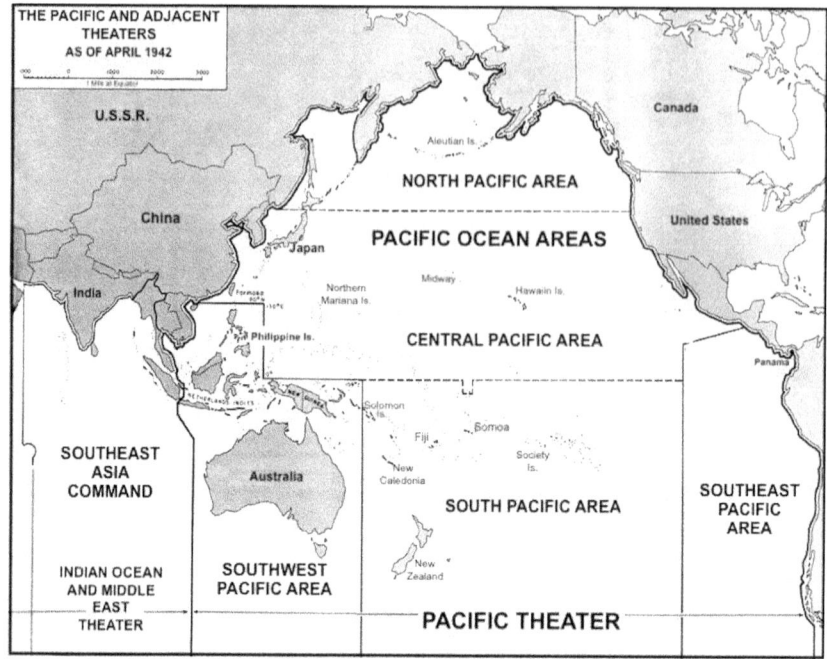

This map of the Pacific Theater and adjacent areas was adapted from map II in Louis Morton's *Strategy and Command: The First Two Years* (*United States Army in World War II* series) (Washington, D.C.: Office of the Chief of Military History, Department of the Army, 1989).

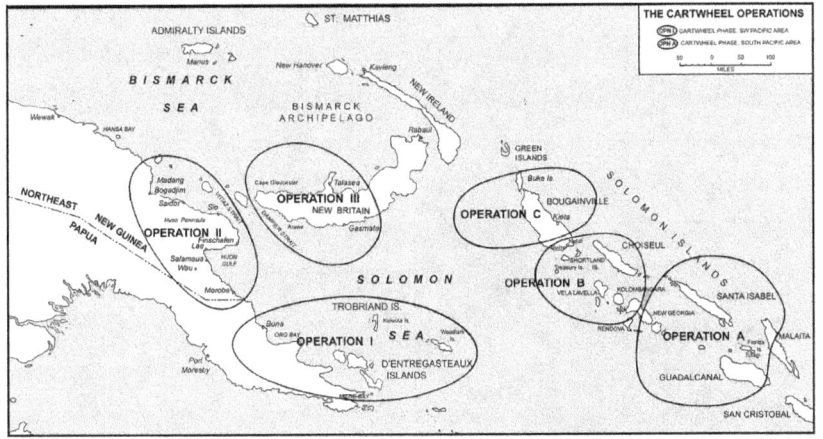

The Cartwheel Operation was designed to surround and neutralize the Japanese base at Rabaul on New Britain. As each island was taken, airfields were built or repaired to serve as new bases from which to attack the Japanese stronghold at Rabaul. Finally, the encirclement of New Britain left the Japanese troops at Rabaul isolated without air or naval protection. It was bypassed using the "island hopping" tactics of the Allied forces. Map redrawn from map III in Louis Morton's *United States Army in World War II: The War in the Pacific Strategy and Command: The First Two Years* (Washington, D.C.: Center of Military History United States Army, 1989).

CONTENTS

PREFACE

"Today, you honor a man who lived his life with patience and purpose. His children will be able to tell their children that he not only stood for all that is good, he always took action."

– U.S. Representative John D. Dingell Jr. (D-MI), May 24, 2011

"HERE HE COMES!" Stephen shouted from his gunboat's conning tower the moment he saw the kamikaze headed directly toward the heart of his ship.

Twenty-year-old Stephen Ganzberger's eyes were glued to his binoculars on that humid, tropical October morning in Leyte Gulf, Philippines. "I was watching through the binoculars, and at the time I was steering the ship," he explained, "and I kept watching him. I kept looking at him . . . and pretty soon you could see that airplane was coming exactly where [I] was standing."

———

I'll never forget the rain. Nearly sixty-seven years later, I stood there half soaked with my family and a group of Marines. Arlington National Cemetery. The most sacred place where our country's fallen heroes go to rest. In my right arm I cradled an urn. The front of the urn held a picture of a young sailor standing on a sandy island in front of a tree. The picture had a tiny autograph written on it. The words "Love Stevie" were delicately penned on the bottom right-hand corner. A Marine and the Arlington staff began escorting me to Stevie's final resting place. I gripped the urn and held it as tightly to myself as I could. As I walked under an umbrella, the rain around me created a mood that felt like the angels themselves were weeping for the loss of such a great man. It was September 7, 2011.

I was never supposed to have carried Stevie's urn. That duty had always been reserved for his daughter, my Aunt Barrie, his youngest child. Like every father's youngest child, Barrie held a special place in his heart, and he in hers. But once my family had arrived in the main waiting room of Arlington before his funeral began, my aunt approached me and told me she wanted me to carry his urn to his final resting place. She mentioned that he would have wanted it that way. I will always be eternally grateful to my aunt for doing that, especially because I know she loves him as much as I do. It meant the world to me.

But there was something else eating away at me on that dreary day. For as lucky as I was to have known my grandpa, I felt robbed at the same time. This wasn't how it was supposed to end. My grandpa and I were supposed to have sat down and spent the entire summer of 2011 recording his World War II story together. After decades of silence, he wanted to spill every last detail. One of the last things he ever said to me was, "we're going to go back," referring to the start of his story, "and then see what I can fill in." Then he said, "We're gonna be all summer long on this. I'll make sure we fill it in all right." But two days later, he was taken from me. Sadly, as is life, it was yet another opportunity just barely missed. The reality of quite possibly never uncovering his story after so long was too much to bear.

But there was hope. There were bits and pieces of information I had gotten from him over the years, glimpses into his wartime past that I'd collected. This included an interview I'd gotten just days before he died, as well as a home movie from 1997 that my uncle sent me. I'd also obtained some of his official records from the National Archives Personnel Center. As an avid World War II historian, I was determined to get the rest of his story no matter what. Something deep down within me made me feel like I had no choice. I just had to. Through the rain, I continued walking with my grandpa on his final journey. Once at his grave, I stopped and placed the urn inside the vault and said my final goodbye. At the moment I laid his ashes to rest in Arlington, I began another journey entirely.

It all started three months earlier in a small kitchen in Southgate, Michigan.

On May 18, 2011, I sat down with my grandpa at his home and interviewed him about his World War II experiences in the Pacific Theater. He served in the U.S. Navy from 1942 to 1945, and as he spoke, I recorded it all. We spent the afternoon discussing things he hadn't spoken to anyone about in more than 65 years, including how and why he was awarded his silver battle star and two bronze battle stars. After more than two hours, we had managed to gloss over the basic gist of everything, but most of the details were missing. The story was far from over and we both knew it, so before I left, we made each other a promise that we would spend the entire summer pouring over the details on each of the seven campaigns he was in. It was one of the most special days of my life. I had no idea it would be the last time I would see him conscious and alive.

He passed away two days after our interview on May 20. He was eighty-six.

I am extremely blessed and fortunate to have known him and been able to call him my grandfather. The indelible mark he has left on me is something I am proud and privileged to carry with me for the rest of my life. He was and to this day remains my hero. I would not be the man I am if not for him.

With the limited amount of information that I did get from him, I was determined to find resources that could help fill in the details of what happened to him and his shipmates during the war. What really happened during the "beer stealing" incident? How exactly did his close buddy die? What was he talking about when he claimed his skipper abandoned his fellow Americans in the water on Bloody Sunday? I knew there was more. So much more. More, it turns out, than I could have ever possibly imagined.

From talking to my grandpa, I knew he served aboard something called a Landing Craft Infantry (LCI). I decided to search for his shipmates, hoping there were still some alive. I began in September 2012, at the USS LCI National Association annual reunion in Charleston, South Carolina, where I met dozens of Navy veterans who served aboard LCIs in World War II. That year the LCI Association held a joint reunion with the veterans of LSMs (Landing Ship Medium). Sadly, I wasn't able to meet any fellow shipmates from my grandpa's two ships, USS LCI (L) 329 or the LCI (G) 65 in Charleston. I did, however, have the pleasure of meeting and speaking with several Navy veterans from my grandpa's LCI flotilla group. Since some of them fought with my grandpa's ships, they began explaining details to me. Slowly, other veterans from both theaters of war gathered around and began retelling LCI sea stories. I was hooked. I came for details about my grandpa but left the reunion realizing I wanted to hear all their stories.

While there I was able to meet all sorts of incredible individuals involved with the association, including the ones who helped organize the annual reunions. The LCI veterans I met there became close friends, and I kept in regular contact with one of the last living ones who was still with us up until the completion of this book in 2021. Nearly all of them, sadly, have passed away in the nine years since. I will always cherish the memories I have with those fine men and consider myself lucky to have earned their friendship and trust for them to tell me their stories. But the man who truly stood out to me the most was the president of the association, a wonderful and lively writer named John Cummer. He served as a gunner's mate aboard the LCI (L) 502 during the D-Day invasion at Normandy, France (Gold Beach), on June 6, 1944. Though I was completely unaware at the time, John would become a mentor, close friend, inspiration, and second grandfather to me. It was John, a fellow writer himself, who offered me a chance to get involved by editing the association's quarterly magazine newsletter called the *Elsie Item*. I didn't even need a moment to think. I immediately jumped at the opportunity. It would mean I'd get a first-hand look at all the LCI veterans' stories before they were published. Perhaps I'd eventually come across my grandpa's shipmates, I thought.

In the years since then, I have interviewed dozens and dozens of LCI veterans to hear their stories of what life was like serving aboard an LCI vessel during World War II, in both Europe and the Pacific. In 2013 I began writing articles for the *Elsie Item*, based on my interviews as well as other memoirs I received in the mail from the LCI vets. By the end of 2013, as health issues became more frequent in John's life, he offered me his position of editor-in-chief of the *Elsie Item*, and I accepted. Through my work with the LCI Association, I finally connected with one of my grandpa's shipmates, J.R. Reid, in 2014. As my journey continued as editor, I learned more about my grandpa's story. So much of what I discovered was beyond belief.

I've spent the last decade gathering material from LCI veterans with the purpose of completing my grandpa's story. In that time, I have received hundreds of World War II stories, articles, documents, books, memoirs, and photographs. Aside from the dozens of personal interviews I've conducted with LCI veterans, I've spoken at length with veterans from all branches of the United States Armed Forces who served in World War II. I have collected and sifted through hundreds of pages of official records and documents from the U.S. National Archives that are relevant. I've also incorporated stories from past *Elsie Item* newsletters where applicable and have tried my best to reproduce as many of these LCI veterans' stories as accurately and as best I can throughout my grandpa's story. To ensure the highest quality, countless hours of diligent fact checking and validating all stories to the best of my ability has gone into this book, but as an imperfect historian I also recognize and acknowledge that mistakes can be made. Any inaccuracies or inconsistencies in this book are solely my responsibility alone.

It should be noted that throughout this entire book, the official deck logs—the day-to-day ship's records of the LCI 329 and LCI 65— were used extensively as the principal resource. Passages from the official deck logs were in many cases lifted word for word. The citations for those deck logs can be found in the "Notes" and "Sources" at the end of this book.[1] The author personally finds it important to emphasize to readers that omissions of citations from the deck logs were done for the ease of the reader, not with the intent of passing those words off as the author's own. The best of efforts was made to cite official documents as appropriately and as often as possible, as they were heavily relied upon. This book would not have been possible without the hardworking employees at the U.S. National Archives. All times mentioned throughout the story are in local time (time zones in which they occurred) based on the time zones recorded in the deck logs. As Stephen's LCIs were constantly on the move, it may not be mentioned each time Stephen's ship crossed into a new time zone. A map of all world time zones can be used as a guide on the last page of this Preface. It is also worth noting that all veterans' oral interviews and written material were published in their original language. This includes grammatical errors and certain terms that would likely be found distasteful and frowned upon today. This was done purposely to preserve authenticity.

My hope is that by writing this book, my family and my generation will have the opportunity to acknowledge and treasure the stories I was fortunate enough to obtain from some of these remarkable men from the Greatest Generation—who sacrificed so much fighting for liberty and freedom, not just for America but for the entire world. The most incredible sacrifice of all, though, was the one the men made for each other. As the old epitaph goes, "Theirs is not to make reply. Theirs is not to reason why. Theirs is but to do and die."

These are the men of the LCI. And these are their tales.

Zach S. Morris, Author
Stephen Ganzberger's grandson

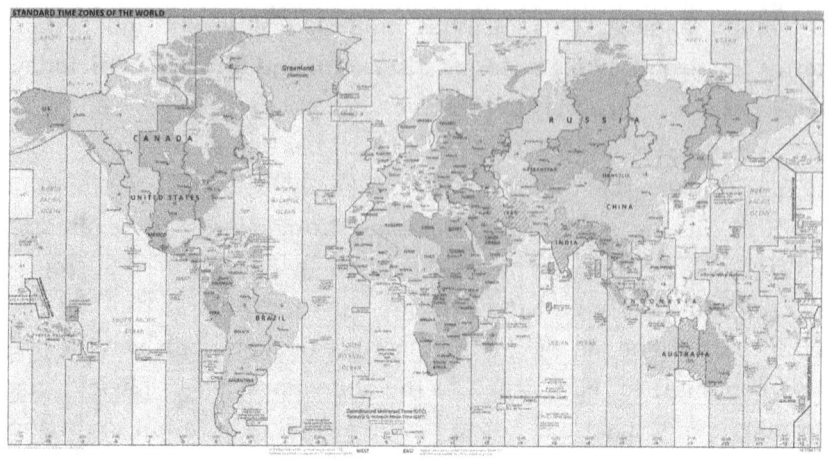

Time zone map. (U.S. Central Intelligence Agency photo)

INTRODUCTION

"Putting the two movements together, landing from the ocean and then fighting on land, was called amphibious warfare."

– Admiral John H. Morrill, *The Cincinnati*
Commander of LCI "Black Cat" Flotilla Thirteen

Soldiers from the 32nd "Red Arrow" Division's 126th Infantry Regiment rush from USS LCI (L) 73 during the invasion of Saidor, New Guinea, on January 2, 1944. (U.S. Army Signal Corps photo, U.S. National Archives)

My grandpa, Stephen Ganzberger, who will henceforth be referred to simply as Stephen, had just turned seventeen years old when the Japanese attacked Pearl Harbor, spurring the United States to enter World War II. Stephen enlisted in the U.S. Navy on August 26, 1942, the day after his eighteenth birthday. Ernie Pyle, one of America's most famous war correspondents, stated that during the Second World War, "The United States Navy had a whole job of embarking, transporting, projecting and landing American invasion troops . . . and afterward keeping the tremendously vital supplies and reinforcements flowing in steadily."[1] As part of that massive Navy, Stephen was selected to be a sailor aboard a Landing Craft Infantry (LCI) in the Pacific Theater. Of all the sins a ship's commander can commit, few are worse than running his vessel aground—an offense so serious it could end an officer's career.[2] However, running aground was the LCI's primary purpose, and because of this, LCIs were utterly lacking in glamour and were universally regarded as the least desirable service the Navy could offer. According to Charles Uhl, LCIs were "scorned by the rest of the navy. . . . After all, we ran our ships aground! And we did it deliberately!"[3]

LCIs were flat-bottomed amphibious assault ships that were used to quickly deliver large numbers of fighting troops and equipment onto enemy beaches in both the Pacific and European Theaters in World War II. LCIs were so small the U.S. Navy did not even give them names, only numbers to designate them apart, which made some of their young crewmen feel embarrassed that they only had a number and not a name. The ships were a specialized, unique, and critical component of the newly created Naval Amphibious Forces. This uniqueness had to do with their miniature size. LCIs were only 158 feet long and 23 feet wide, much smaller than traditional Navy vessels such as battleships. They had permanent crews of 24–60 sailors and were able to carry an additional two hundred troops, who descended from ramps on each side of the craft upon beaching. LCIs were able to quickly get off the beach after shifting into reverse, as veteran Robert W. Kirsch explained because, "On the way into the beach, a stern (rear) anchor would be dropped," and since the LCI would be lighter after the troops left the ship, he said the rear anchor would then be "used to pull the LCI off of the beach."[4]

LCIs were never designed for cross-ocean travel—in fact, added Kirsch, "It was called a craft rather than a ship as it was thought that it could not traverse large bodies of water." Original plans called for LCIs to be shipped over to their theater of war in large pieces and assembled upon arrival. Yet given the urgency of wartime, somehow those little LCIs miraculously crossed the oceans all on their own, sailing from the United States to the European and Pacific Theaters at a maximum speed of sixteen knots. Unfortunately for their crews, the flat-bottom hulls of the LCIs were designed for beaching—not slicing smoothly through the water—which meant when they sailed, they felt every wave. An admiral once famously dubbed them "water-bugs" due to their small size and maneuverability. LCIs were commonly referred to by their most popular nickname "Elsie Items,"

which comes from their initials in the Navy's phonetic alphabet—
L.C. Item—L.C.I.

There were three primary roles that LCIs mainly served during
the war. As discussed, the first and most important role in the beginning was landing up to two hundred fully equipped troops and their
supplies on an enemy-held beach in a single trip. The second was
providing smoke screen cover for larger ships in the convoy, which
was done to complicate visibility for enemies that attempted to attack
the larger Allied ships. This role saved many lives aboard ships
targeted by Japanese kamikaze planes, especially in the Philippines
and Okinawa campaigns. The third role, which slowly became the
primary role of LCIs toward the end of the war, was providing direct
covering fire for the infantry as they unloaded onto the beaches
during the landings. This third role expanded even more dynami-
cally as the LCIs were converted into gunboats, rocket ships, and
mortar ships in late 1943 and throughout 1944. The success of the LCI
in this third role even led directly to the creation of its big brother
later in the war, the Landing Craft Support (LCS). In addition to their
three main functions, LCI 711 veteran Russell Hartwell wrote, "The
LCI's did everything from making beach-head landings to hauling
ammunition and high-octane gasoline for American fighter planes.
They also rescued people from enemy held territory, downed pilots
and shipwrecked sailors. The LCI provided fire power cover for
landing troops against enemy targets on the beach and the jungle's
edge and against the enemy aircraft." Hartwell went on to say that
LCIs "delivered the wounded to main Army hospitals."[5] LCIs also
transported prisoners of war (POWs) in both theaters. LCIs were
known for their bravery in coming to the rescue of imperiled ships by
helping them fight fires. And in one particularly rare instance aboard
an LCI, the crew even delivered a baby girl birthed from an Italian
refugee.[6]

It was realized that smaller ships like LCIs were needed to cover
the Allied landing troops when the heavy shelling from the larger
American ships like battleships ceased firing on the beaches immediately before an assault. Once the battleships stopped bombarding the

beach it created a dangerous lull period right up to the moment the vulnerable troops came ashore. This gave the Japanese defending the beaches just enough time to regroup and reorganize, where they would frantically use these precious few minutes to open back up their bunkers, caves, and pillboxes and reposition their artillery and guns to fire on the invading Allied troops. This situation called for a small ship that could quickly transport many troops to the beach while providing suppressive covering fire for the amphibious tractors (also called Landing Vehicle Tracked, or LVTs) and landing boats making the initial assaults at the same time. LCIs were the answer.

All of this meant that LCIs had a front row seat to the horrors of war as they protected the Allied landing assaults during the first waves of invasions.

———

Stephen's LCI group was assigned to the South Pacific Area in 1943 where he served aboard his first ship, USS LCI (L) 329.[7] Stephen would see his first action of the war aboard this LCI at Rendova Harbor, New Georgia, in July 1943. He would be a part of the island-hopping campaign in the Solomon Islands codenamed "Operation Cartwheel." Cartwheel was a two-pronged, simultaneous attack on the Solomon Islands and New Guinea. General Douglas MacArthur made two things abundantly clear in the early planning stages of the war: the U.S. must retake the Philippines and the mighty Japanese base on Rabaul, New Britain, must be destroyed.

The U.S. would methodically fight their way, island-by-island, starting in the Solomon Islands (Guadalcanal and Florida Island) and work their way northwest to the island of New Britain, which housed the powerful Japanese air base of Rabaul. To achieve ultimate victory in the Pacific, the Allies needed to neutralize Rabaul at the northeast tip of the island.[8]

In January 1944, Quartermaster 3/c (Third Class) Stephen Ganzberger transferred to the LCI (L) 65, because they needed a quartermaster—a navigator of the ship. His LCI flotilla group spent the summer of 1944 supplying, patrolling, and conducting training exercises on various islands in the Southwest Pacific. They'd also spend the first half of 1944 being slowly converted from a Landing Craft Infantry (Large), or LCI (L), into a more heavily armed LCI gunboat, or LCI (G). The LCI (L) 65 would officially be renamed LCI (G) 65 on June 20, 1944, and a month later it was reassigned to MacArthur's Seventh Fleet. Each converted LCI in the 65's group had both their ramps removed. This meant that a large number of LCIs would no longer be involved in landing troops on the beaches. Some LCIs were converted into LCI rocket ships, or LCI (R)s, in which rocket launcher racks used like bazookas were installed on each side of the ship where their ramps had been. Other LCIs were converted into LCI mortar ships, or LCI (M)s, in which they were modified to carry three 4.2-inch heavy mortars, which were projectile explosives that could be lobbed over distant hills as the LCIs approached the beaches.[9] The LCI (G) 65, officially a gunboat in summer 1944, had one three-inch 50 caliber heavy gun, one single barrel 40mm Bofor automatic gun, four 20mm Oerlikon machine guns, and six .50 caliber machine guns. Stephen's ship would be equipped to provide greater covering fire for the landing forces invading the beaches.

In September 1944, his LCI group saw their first action as newly converted gunboats, rocket ships, and mortar ships in the Battle of Morotai. Then in October 1944, the LCI 65 began a three-month-long fight to retake the Philippine Islands of Leyte and Luzon. That's when Stephen's war would begin to take its deadliest turn.

It was the day the kamikazes began falling from the sky.

Stephen Ganzberger in World War II. He served aboard the USS LCI (L) 329 and USS LCI (G) 65 in the Pacific Theater. (Ganzberger Family Collection)

1

A DATE WHICH WILL LIVE IN INFAMY

*"Whatever is asked of us, I am sure we can accomplish it; we are the free
and unconquerable people of the United States of America."*

– Eleanor Roosevelt (radio address, December 7, 1941)

Stephen Ganzberger (middle) poses with two of his brothers, Frank (left) and John, date unknown. Frank served in the merchant marines and John served in the U.S. Army during World War II. (Ganzberger Family Collection)

Stephen was born to parents of Hungarian descent in 1924. His father, Jozsef Gansperger, was born near Budapest. In 1910, Jozsef left Europe destined for the United States aboard the RMS *Carpathia*—two years before the ship became famous for rescuing survivors of the RMS *Titanic*, which had sunk in the freezing waters of the North Atlantic Ocean. Quite possibly, the slip of an official's pen, or more likely, an intentional adjustment made at some point to please the eye, caused Jozsef's name to be rewritten and changed to "Joseph." It should also be noted that the original spelling of the family name, Gansperger, was changed to "Ganzberger" sometime around the early 1940s.[1] Joseph arrived in America the same year a woman named Amelia Geleta gave birth to her first child, Ida, back in Hungary. When Ida's father—Amelia's first husband—died unexpectedly, Amelia and Ida emigrated to America in the years that followed.

The Roaring Twenties began with Amelia having a second child,

William (or Bill), with a second man in 1920, shortly after Joseph had returned to America from the First World War. Two years later, Amelia met Joseph in Wyandotte, Michigan and they married soon after. In 1922, Amelia's third child, Frank, was born. Frank would be the first of four children Amelia would eventually have with Joseph. A year later in 1923, John was born. Then, the following year, on August 25, 1924, Amelia gave birth to a fifth child, a baby boy named Stephen. Stephen's little sister Barbara, the last of Amelia and Joseph's children, would be born in 1927.

Soon after the stock market crash of 1929, Joseph and Amelia weathered the hard times of the Great Depression in the Thirties with their six children. Stephen would tell stories years later about his family's life during the Depression. His memories included hungry days and in one instance he remembered only receiving a single orange as a present one Christmas. The decade that followed was gripped by fascist leaders that took rise all over the world, and it wasn't long before Stephen and his brothers were thrust into war.

———

It was a chilly Sunday afternoon in December 1941. Stephen had just arrived home from church with his family when the news reached the Ganzberger house. Breaking news reports were urgently cutting into regularly scheduled programs shortly after 2:30 p.m. in Wyandotte. The news was shocking. The Japanese had intentionally, and brutally, attacked American forces at Pearl Harbor, who were caught completely by surprise. Many young sailors were still asleep in their bunks aboard the various battleships when the first bombs fell. Stephen had just turned seventeen a little over three months prior.

The Imperial Japanese had made what would later be argued as their most ill-fated military decision in history. Pilots of the Japanese air force, wearing white *hachimaki* cloths marked with the symbol of the rising sun and the legend "Sure Victory" tied around their heads, took off from aircraft carriers in the early morning hours of December 7, 1941, and headed south for Hawaii.[2] Their target was the

American naval base at Pearl Harbor, Oahu, Hawaii. Two waves of
Japanese planes, navigated and directed by the signals coming from
songs broadcasted from Hawaiian radio stations on Oahu, carried out
a deliberate attack on American servicemen and civilians. Despite
the loss of life and devastating damage to the American Navy, the
Japanese bombs and bullets were only victorious at violently jarring
awake a metaphorical giant from its tumultuous slumber.

———

Decades after World War II, Stephen and future Congressman John
D. Dingell Jr. would become close friends through political connections. John Dingell Jr. was an extraordinary man who lived a remarkable life. Eventually, Dingell would go on to serve an impressive fifty-
nine years and twenty-one days in the U.S. House of Representatives.
To this day, that remains the record for continuous service in the
United States Congress.

But at the time of the Japanese attack on December 7, 1941,
Dingell was only fifteen years old and working as a congressional
page.[3] John's father, John D. Dingell Sr., was serving as the Representative for Michigan's 15th District and had appointed his son as a
House page back in 1938. As such, John Jr. helped his father around
the office and ran errands wherever he was needed on the Hill.
Because of this, fifteen-year-old John Jr. was present in the House of
Representatives for President Franklin Roosevelt's famous speech
that was delivered to a jam-packed House the day after the attack on
Monday, December 8. In his 2018 book, *The Dean*, John Dingell
described the historic events of those two days:

> My ten-year-old brother, Jim, and I were walking home from Mass
> at St. Joseph's Church. . . . A man we didn't know ran up to us and
> shouted, "The Japs just bombed Pearl Harbor!" . . . Jim and I sat
> glued to the radio for the rest of the afternoon. . . . Unlike most
> Americans, Jim and I knew exactly where Pearl Harbor was. We had

stayed there with our dad about five years earlier, on one of his congressional trips. . . .

I got up early the next morning fueled by my dad's anger, and my own. Large crowds had been building up outside the Capitol Building all night. Soldiers with .50 caliber machine guns were stationed on the roofs of government buildings including the Capitol. A perimeter fence had been erected hurriedly overnight to keep civilians a safe distance away. . . .

As I made my way up the Gallery . . . I heard the sharp crack of the Speaker's gavel. [Speaker Sam] Rayburn called on all unauthorized people to give up their seats for the Senators who'd soon be arriving. People gathered behind the railings on the floor as a clamor rose again and the Senate started to file in.

The five hundred seats in the visitor and press galleries were already full to overflowing. . . . I moved carefully through the crowd. . . . People were too excited to sit. . . . I paused for a moment when I caught my first glance of the full extent of the floor: . . . wives and children were everywhere. This was not like anything I had ever seen before in the chamber. . . .

Microphones were secured on the rostrum where the president would speak. Dozens of newsreel cameras were pointing toward the still-empty spot. . . . Rayburn's gavel sounded again, and the room fell silent. By now, no space was unoccupied. People stood on chairs, sofas, and even the paper-thin wall panels. . . .

When the president was announced, the entire chamber rose as one to greet him with a standing ovation. . . . The president's son, James, who was wearing the formal, blue dress uniform of the U.S. Marine Corps, held firmly on to his father's other arm. . . . I thought of the courage it took for FDR to rally himself, polio be damned, and literally rise to the occasion."[4]

President Roosevelt, standing front and center, then began his speech that started with the famous line, "Yesterday, December 7, 1941 —a date which will live in infamy—the United States of America was suddenly and deliberately attacked by naval and air forces of the Empire of Japan. . ."

A captivated nation listened intently to the President's every word live over the radio. Immediately after Roosevelt's speech, Dingell described the feeling of such a historic moment:

> Overwhelming applause and cheers followed Roosevelt as he turned and left the microphone. All around, you could see unashamed tears streaming down the faces of strong men in the chamber. Watching this extraordinary display of open emotion from adults was something I had never seen before. . . .
>
> FDR gave the American people the unvarnished truth: we were attacked suddenly and deliberately by a ruthless force that now had to be defeated. . . . Even more important, he had called us to the collective national belief that we could, we *must*, win the war. . . . on December 8, 1941, what I knew for certain was that I had just witnessed our republic at one of its finest hours.[5]

Like Stephen, Dingell would have to wait until he turned eighteen before he could enlist in the armed forces. But unlike Stephen, Dingell would have to wait slightly longer, as he could not enlist in the U.S. Army until his eighteenth birthday in July 1944.

Students and teachers of the Capitol Page School in 1939. John D. Dingell Jr. is in the front row, ninth from the right. (Courtesy of the Capital Photo Service, Collection of the U.S. House of Representatives, Jim Oliver Collection)

From the knowledge we possess at the present day, there are multiple reasons behind why the Japanese unleashed the unprovoked, ferocious attack that morning.

Author John D. Lukacs explained one reason saying, "According to Japan's militarists, the rising sun of Amaterasu, the ancient goddess of creation, was waking the master Yamato race to its destiny. The annexations of Formosa, Korea, and Manchuria, followed by an invasion of China in 1937, signaled Japan's desire to resurrect the holy mission of Jimmu Tenno—Japan's first emperor circa 660 B.C.—called *hakko ichiu,* meaning to forcefully bring 'the eight corners of the world under one roof.'" This sentiment was eerily similar to Nazi racial philosopher Alfred Rosenberg's anti-Semitic wishes that Germany's "master race" of Aryans be protected against supposed racial threats from the Jewish people in Europe, which had been a motivating factor behind Adolf Hitler's attempts to conquer Europe. Historians say this emerging form of manifest destiny was the reasoning behind the Imperial Japanese setting out to conquer the Pacific in the 1930s and 1940s. At the outset of World War II, the Japanese claimed control over land that extended from the Philippines, Formosa (present-day Taiwan), much of New Guinea, Micronesia, the Gilbert Islands, the Solomon Islands, and Southeast Asia, all the way north to the Alaskan Aleutian Islands. And it was all carried out in accordance with Japan's strict Bushido samurai warrior code—in which all Japanese personnel had been indoctrinated. Like the ancient Greek Spartans, the paramount goal of life was a glorious death on the battlefield, with surrender being the ultimate dishonor. In Lukacs's words, "One who surrendered, therefore, defied his destiny of death, betrayed his emperor, his country, his family, and his comrades. In essence, surrendering betrayed *yamamoto damashii* —the very soul of Japan."[6]

As another contributing factor of Japan's aggression, many have pointed to the massive blockade Americans had placed on Japan's forces due to Japan's aggression in what was seen around the world as an illegal invasion of Manchuria and China. In the years leading up to the Pearl Harbor attack, Roosevelt had halted exports of oil, rubber, and iron to Japan. He also froze Japanese assets in the U.S. and prevented Japanese merchant vessels from using the Panama Canal. Historians have often cited the passage of the Lend-Lease Act by the U.S. Congress as another major provocation to the Axis powers.

Author John Koster wrote an interesting and impressively well-researched book in 2012 titled, *Operation Snow*, which details how a Soviet spy named Harry Dexter White heavily influenced several of Roosevelt's key decisions on Japanese foreign policy from within FDR's own cabinet in hopes of provoking a war between the U.S. and Japan—essentially preventing a Japanese invasion of the U.S.S.R. from the far east. The Soviets were aware they could not win a fight on two fronts against both Germany and Japan simultaneously. Perhaps because White succeeded, the Soviets deserve a far larger portion of the blame for orchestrating a war between the Japanese and the U.S.[7]

But regardless of the reasons behind the devastating attack on Pearl Harbor, over twenty-four hundred Americans lost their lives that December day. The course of human history was altered forever. America was officially at war.

———

Upon America's declaration of war against Japan, Hitler declared war against the Unites States. Three days later, on December 11, congress responded and voted to declare war on Nazi Germany. From that moment, and for the next four years, millions of American men and women poured into factories and shipyards to assemble the greatest naval fleet the world has ever seen. On Christmas 1941, the U.S. Navy was ill-prepared for war, especially in the desperately needed areas of ship-to-shore landing craft, but that would soon change. For it had been the British who first realized upon the fall of France and the disaster at Dunkirk that the war against Hitler would have to be an amphibious war, as the French ports were no longer available—at least not until they could be assaulted and recaptured. According to John C. Reilly Jr., the challenge was that "Entire armies, not small raiding forces, would have to be transported and landed. The old scheme of debarking troops from transports offshore and landing them in small davit-carried craft would not suffice; something more was needed."[8]

Stephen wanted to enlist in the armed forces the morning after the attack. However, at that time, a young man needed permission from his parents to enlist if he was younger than eighteen years old. Unfortunately, Joseph and Amelia forbid their sons from enlisting until they were legally of age. Stephen's half brother Bill was twenty-one and soon enlisted in the Army. His older brothers Frank and John were old enough to join the merchant marines and Army, respectively. But Stephen was only seventeen and the only one of his brothers not old enough to enlist. He would have to wait more than eight long months before he could join the fight.

All the while, a genius architect named John C. Niedermair was helping the U.S. Navy design new ships for the coming war.

2

AUGUST 1942–JUNE 1943: ENLISTMENT AND JOURNEY ACROSS THE PACIFIC

"The LCI is a bastard craft. Its initials mean, 'landing craft infantry,' but these words do not tell the story. It's bigger than a Higgins boat, and smaller than a minesweeper and it crosses oceans under its own power, rolling until its flag dips water. Designed to land troops on shelving shores, the LCI has a flat bottom and draws only a few feet of water. It is square-sterned and blunt-browed. It has the beauty of a freight train caboose and the wallop of a locomotive. At sea its officers and crew are sometimes too seasick to care. In attack, their morale is as great as their ships are small."

– Richard W. Johnston, United Press[1]

Stephen celebrated his eighteenth birthday on August 25, 1942. The next morning, he waited in line to enlist the minute the Navy recruiting station opened. His World War II journey into the hellish Pacific Theater was about to begin.

Throughout the summer of 1942 while Stephen patiently waited to enlist, the U.S. Navy high command had already begun testing all their new different types of amphibious landing craft up and down the Atlantic coast. One of the most notorious and often-cited stories in LCI lore comes from Vice Admiral (then Captain) Daniel E. "Uncle Dan" Barbey, known to military historians as the "father of the Amphibious Forces," as he had been among those who spearheaded its creation and development even before World War II officially broke out. In his history, Barbey spoke about a particular amusing incident involving LCIs:

> Trying out each new type of amphibious craft and vehicle before it went on the production line was one of my jobs during the summer of 1942. Some of these try-out jobs had their interesting moments. One Sunday afternoon off the Cape Cod coast, on a trial run of a new LCI, we asked the builder's representative to run it up on a beach and not to be too cautious. He wasn't. The craft stopped only after it had crossed a scenic highway and disrupted traffic. But the test was useful. It gave us confidence in the ability of the ship to take heavy punishment.[2]

Men like Barbey were extremely proud to be part of the Naval Amphibious Forces (A.F.). However, not everyone was thrilled to be assigned to the A.F. during World War II. In fact, the A.F. had quite a rancid reputation in the Navy. The men relegated to the A.F. were made to feel inferior, like outcasts from society. During the first years of the war, they used to openly refer to themselves as "we peons." Your typical big-ship-type Navy officers, who could hardly imagine going up on the beach on purpose, avoided LCIs whenever possible. According to the first volume of *USS LCI Landing Craft Infantry*:

> High officials in particular looked upon the Amphibious Force as the Siberia of the Navy, a place of exile. This belief was fostered by the questionable practice—during 1942 and 1943—of "dumping" senior officers "passed over" for promotion, onto AFAF organiza-

tions. And it seemed that everywhere there was disorganization where there should have been order. Probably no lower morale will ever be experienced by United States Navy personnel.[3]

The A.F. had such an awful reputation that Commodore John H. Morrill—who would go on to command "Black Cat" LCI Flotilla Thirteen and later become an Admiral—was even advised against volunteering for the A.F. in November 1943. A graduate of the Annapolis Naval Academy, Morrill was yearning for action after his harrowing escape from the Philippines in early 1942. He had specifically requested duty aboard a landing craft infantry at the Navy's Bureau of Personnel, because rumor had it there was combat action. Morrill remembered that his old classmate and buddy, B. L. Austin, the Detail Officer, was incredulous and horrified when Morrill requested command of a flotilla of LCIs. As Morrill remembered, Austin replied, "'You are asking for a basket full of trouble, John. Are you sure you want that kind of duty? The people in the Amphibious Forces are not like those we grew up with. . . . It would be a dead end for you, John, no promotion for you, only trouble. . . . I don't think you can do much with that kind of people.'"[4]

———

Like most of the young men who enlisted in the Navy during World War II, Stephen's training started off at the U.S. Naval Training Station in Great Lakes, Illinois.

"That was a pretty fast time," Stephen remembered. "I enlisted August 26, 1942."

One of only several photographs he possessed from World War II was taken on September 15, only three weeks after he enlisted. It was a large group photo from basic training of one hundred and sixteen sailors, including the commander, standing at attention. He pointed to it as he continued, "and by September fifteenth, you see the remnants of it. [We shipped out] about a week after that, but prior to that we was exercising every day, you know? Just getting your body ready for coming events." When his unit completed basic training, they headed down to an amphibious training base in Portsmouth, Virginia. After several days they headed to Texas.

CO.831 "42" R.E. HAYNES – C.SP. CO. COMD. SEP 15, 1942 U.S. NAVAL TRAINING STATION-GREAT LAKES, ILL. Sailors pose for a photo in boot camp at the training station in Great Lakes, Illinois on September 15, 1942. Taken only twenty days after enlisting in the United States Navy, Apprentice Seaman Stephen Ganzberger can be seen in this photograph standing in the 3rd row, 11th from the right. (Ganzberger Family Collection)

"We stayed at a U.S.O. until we got our ship ready to go down to Brown City, Texas. That's where Brown Shipbuilding Company is down there. That's where my ship was." Stephen continued: "When we left Brown City, we went down to Galveston. And there, we went out two or three times into the Gulf of Mexico on a shakedown, to see if the ship was operational. And I was on midnight watch." Stephen explained that a shakedown cruise was when sailors tested out their ship for the first time after it was built and commissioned to learn how to handle everything and make sure it functioned properly. As Stephen described it in 1997, "A shakedown is like . . . when you buy a new car, you got to drive it for a while, then when all the kinks get out, then they'll deliver you the car." The shakedown cruise tested the ship for any possible structural or functional weaknesses and was used to train the crew until they became an efficient team.[5] But a shakedown cruise could be nerve-wracking at times. John Morrill remembered the hideous noises his newly constructed LCI made as it loudly creaked and groaned during its shakedown cruise on the first night out of Norfolk. In his book Morrill wrote, "One of the young crewmen on the bridge asked, 'Commander, do you think that the women welders that put these ships together did all right?' I realized that he was actually fearful that the ship might break apart, so I gave him the age old reply, 'Young man, what you feel and hear is the sweetest music to a sailor's ear. It is a good ship coming alive and testing its own strength.'"[6] However, on many LCIs the entire crew, including the officers, was without experience. Chuck Savard of the LCI 448 recalled that on their shakedown cruise, "We headed into the Atlantic with a 100% green crew. When we were a few miles out the skipper said, 'What do I do now?'"[7]

———

Among the many different drills the crews practiced during their shakedown cruises were fire drills, abandon-ship drills, collision drills, beaching and retracting, and perhaps the most important drill, practicing battle stations—or "General Quarters." In his book *Brave Men*, Ernie Pyle wrote, "General Quarters is the Navy term for full alert and means that everybody is on full duty until the crisis ends. It may be twenty minutes or it may be forty-eight hours." Aboard any ship in the navy, a sailor was under one of three conditions at all times, meaning there were three conditions of battle stations. Condition I was General Quarters. Under Condition I, all battle stations and guns were manned, and usually surface or air action was imminent or about to take place. At General Quarters, the men were at their battle stations ready to fight. Each man was at his designated battle station—and exactly where a sailor's battle station was chosen heavily depended on each sailor's specific rating, which will be discussed shortly. Condition I was sometimes modified to allow a few persons to rest at a time on station or to let designated personnel, like cooks and bakers, draw rations for delivery to battle stations. Condition II was a special watch used by gunfire support ships for situations such as extended periods of shore bombardment. According to Ernie Pyle, "Condition Two is half alert, four hours on, four hours off, but the off hours are spent right at the battle station. It merely gives the men a little chance to relax."[8] Condition III was the normal wartime cruising watch. Normally, when cruising under Condition III, the ship's company stood watch on a basis of shifts that were four hours on, eight hours off, while about one-third of the ship's armament was manned in the event of a surprise attack.[9] During General Quarters, Stephen was the gunner assigned to the number-two 20mm machine gun at the center of the ship.

The first LCI built was the USS LCI (L) 209, which was commissioned on October 1, 1942. The second LCI built was the USS LCI (L) 1, commissioned on October 7, 1942. Chronologically from then on, LCIs were manufactured and assigned numbers by shipyard in an unbroken ascending line of order. On November 8, 1942, Stephen's first ship, USS LCI (L) 329, was commissioned for duty. The LCI (L) 329, hereafter referred to as the LCI 329 or simply the 329, was part of the still-being-built LCI (L) Flotilla Five destined for the Pacific Theater. Listed below are the first twenty-six ships assigned to LCI Flotilla Five that were built in four different shipyards and commissioned between October 31 and December 17, 1942 (by yard and commissioning date).[10]

New York Shipbuilding Corp. Camden, NJ
USS LCI (L) 21.........December 8
USS LCI (L) 22.........December 8
USS LCI (L) 23.........December 8
USS LCI (L) 24.........December 8

George Lawley & Sons, Neponset, MA
USS LCI (L) 222.........December 3
USS LCI (L) 223.........December 8

Consolidated Steel Corp. Ltd., Orange, TX
USS LCI (L) 61.........November 12
USS LCI (L) 62.........November 13

USS LCI (L) 63.........November 16
USS LCI (L) 64.........December 12
USS LCI (L) 65.........December 14
USS LCI (L) 66.........December 14
USS LCI (L) 67.........December 17
USS LCI (L) 68.........December 17
USS LCI (L) 69.........December 24
USS LCI (L) 70.........December 24

Brown Shipbuilding Co., Houston, TX
USS LCI (L) 327.........October 31
USS LCI (L) 328.........October 31
USS LCI (L) 329.........November 8
USS LCI (L) 330.........November 9
USS LCI (L) 331.........November 16
USS LCI (L) 332.........November 17
USS LCI (L) 333.........November 24
USS LCI (L) 334.........November 24
USS LCI (L) 335.........November 27
USS LCI (L) 336.........December 3

Note: The East Coast–built ships would rendezvous with their Texas counterparts on February 25, 1943, in the Panama Canal Zone. In the summer of 1944, upon the completion of Operation Cartwheel, elements of LCI Flotilla Five were transferred from the Third Fleet to the command of the Seventh Amphibious Force's LCI Flotilla Fifteen.

———

An LCI flotilla, aka "the flot," totaled thirty-six ships: six LCIs per division, two divisions per group, and three groups per flotilla. The entire flotilla was typically commanded by a captain or ranking commander. Each LCI group comprised twelve ships and was usually under the command of a commander or a lieutenant commander. A division made up of six LCIs was often led by a lieutenant commander.

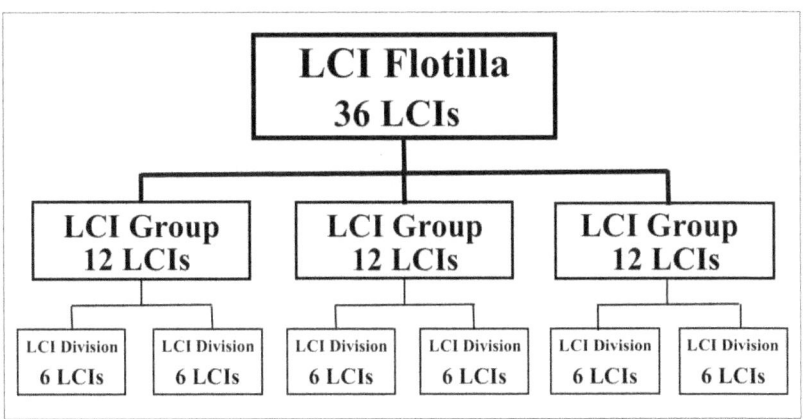

The basic organization of an LCI flotilla, which consisted of thirty-six LCIs.

LCI Flotilla Five Organization:[11]

Commander LCI Flotilla Five – CDR. C.L. Walton

Group Thirteen: LCDR. M.M. Byrd, USN.
• LCIs 21, 22, 61, 62, 63, 64, 65, 66, 67, 68, 69, 70.
Group Fourteen: LCDR. A.V. Jannotta, USNR.
• LCIs 23, 24, 327, 328, 329, 330, 331, 332, 333, 334, 335, 336.
Group Fifteen: CDR. J. McDonald Smith, USN.
• LCIs 222, 223.

LCI 329 would be assigned to LCI Flotilla Five's Group Fourteen under the command of Lieutenant Commander Alfred Vernon Jannotta. Jannotta, a veteran of the First World War, was revered by his men who served under him. The skipper of the LCI (L) 334, Alfred Ormston, had this to say about Group Fourteen's commanding officer:

> We reported to the LCI Flotilla 5 headquarters at the Cotton Docks there and were assigned to Division 28 under Group 14 headed by LCdr Alfred V. Jannotta, a WWI vet and really fine man. He put a bunch of us green officers and men in a new type of ship. Nobody knew how to handle them but we went out and trained anyway.[12]

Ormston also recalled a most endearing memory that the commanding officers of his LCI group shared with Jannotta from Christmas 1942:

> A few lucky crew members with families nearby got to go home for a few days. Commander Jannotta had a Christmas party and presented each one of the LCI skippers in his group with a ship's bell which the Navy Department had failed to put on the LCIs.[13]

Galveston, Texas, would be the LCI 329 crew's final stop in the U.S. before shipping off to war in the Pacific. Some men of the crew, including Stephen, would not see the States again until nearly three years later. It is here in Galveston that Stephen would be joined with men such as Elmo Pucci, who he would ride the waves into battle with during the year of 1943.

––––––

A month before the attack on Pearl Harbor, on the afternoon of November 4, 1941, in response to a British request sent to the U.S. Navy's Bureau of Ships, John C. Niedermair picked up a pencil. He began drawing a sketch and doing some basic calculations. Those moments from when he started to when he finished could be looked at as being among the most important and dramatic moments in the entire history of ship design. Those sketches of an entirely new and radical ship became the Landing Ship Tank (LST)—which would later be known as the workhorse of World War II. At the beginning of the war, Niedermair served as the U.S. Navy's Bureau of Ships' Technical Director of Preliminary Ship Design, and once war broke out, his role was to be the brilliant mind that designed all the ship-to-shore landing craft that the U.S. Navy quickly needed developed that didn't yet exist.

"[I]t wasn't long," Niedermair remembered, "before we got going on other projects." After helping create the LST, he recalled in his oral history, "the next . . . landing craft that was very important was the LCIL, landing craft, infantry." He added that the LCI, "was a very interesting thing to develop."[14]

Born on November 2, 1893, in Union Hill, New Jersey, Niedermair possessed a gifted engineering mind from a young age. Author Mel Barger wrote, Niedermair "was selected for a scholarship to attend Webb Institute of Naval Architecture and Marine Engineering in 1914. He graduated first in his class in 1918." Barger went on to write: "After serving briefly as a Naval Officer candidate, [Niedermair] accepted a permanent appointment in December 1918 as a ship draftsman at the New York Navy Yard and served there until joining the Preliminary Design branch at the Navy's Bureau of Ships, where he would ultimately become the highest-ranking civilian in the Bureau." More than eight thousand ships would be built from designs originated under Niedermair's guidance—a record that is never likely to be equaled. As impressive as his work on the LCI design was, his greatest pride, however, was helping design the LCI's big brother, the LST. According to Barger, "it was said that his role in designing the 'Large Slow Target' brought him the most professional satisfaction of his career." Niedermair would go on to become one of the highest-honored civilians in Naval history, including being presented with the Distinguished Civilian Service Award, the Navy's highest honorary award, in 1945.[15]

"We were adding a bunch of new ships and they kept telling me the steel mills couldn't do it." Niedermair recalled in the early 1940s. As a result, said Niedermair:

In connection with the LST and these other craft, the LCIL, we started using the steel that they rolled for automobiles with the fast rolling mills that they had for that purpose. That steel was not considered very good because often you would get folds in it. The steel would be going through the mills and it would fold under ahead and there would be an overlap in it. The edges of these plates that they made were rather weak and they didn't think they could use the sheet metal at first, but then they finally did use it by trimming the edges of all the plates that were rolled for this sheet stuff.[16]

"You must remember," added LCI 74 veteran Robert Kirsch, "that the skin of a LCI is only a quarter inch thick steel plate." The LCI's design stressed ruggedness and ease of mass production.[17] A single LCI was built every few days, and by February 1943, LCIs were being delivered at a rate of ten a month (per shipyard) to the U.S. Navy.[18]

Once built, Niedermair and the U.S. Navy did everything in their power to get the ships and their assigned crews ready for the ensuing battles as quickly as possible. The officers of the LCIs would be locked in a constant state of struggle when it came to acquiring parts and equipment for their little ships throughout the war, but this was especially true in 1942. Niedermair described the host of challenges faced with designing a ship like the LCI, saying, "We ran into a considerable amount of trouble getting engines for these little LCIs." He described the LCIs' engines saying, "We had to use . . . little gray diesels. We put four of them on each shaft. They had two shafts. We packed them well together and geared them to a shaft, and that's the way we engined the LCILs." But those little gray diesel engines broke down all the time and regularly required repairs and locating replacement parts for them in the South Pacific, or anywhere, proved exceedingly difficult. "We also had trouble getting pipes and plumbing for them." Niedermair added, "The yards and the manufacturers were pretty well saturated in early 1942." But it was "the plumbing," he said that presented a most difficult challenge because "we couldn't get any plumbing to speak of." Meaning LCIs were made with hardly any plumbing at all. However, "It was this sort of short-cutting and cutting of red tape," Niedermair reflected, "that made it possible for us to get all those ships . . . put together."[19] Between 1942 and 1944, a total of 921 LCI(L)s were built in American shipyards.

As 1943 approached, while Niedermair helped other LCIs procure the necessities, the LCI 329 had everything she needed despite all her shortcomings, and soon the rest of the ships in Stephen's LCI flotilla would head to the Panama Canal, en route to the Pacific Theater.

―――――

In January 1943, the Allies found themselves for the first time in a position to go on the offensive in both theaters of war. They had just come off victories in North Africa ("Operation Torch") and on a lima-bean shaped island located in the Solomon Islands in the South Pacific called Guadalcanal. The Battle of Guadalcanal that began with the invasion of the 1st Marine Division on August 7, 1942, three weeks before Stephen enlisted in the Navy, was almost won after nearly six months of brutal fighting. Stephen heard all about the battle from talking to survivors. "That was a tough fight for the Marines on Guadalcanal. They had to really fight," he said. Guadal-canal lies to the southeast of New Guinea and its capture was the first major step in the war against Japan. Codenamed "Operation Watch-tower," the Guadalcanal campaign was the first action by amphibi-ously landed Marines of the Pacific War. The Allies needed to secure an airfield on Guadalcanal, located between Koli and Lunga Point, that the Marines secured the day after landing. Fierce and unthink-ably savage fighting was part of the Marines' lives for the next six months. By the end of the battle in February 1943, Americans had suffered over seven thousand casualties if including deaths from jungle diseases like malaria.[20] After all that bloodshed, Guadalcanal was only the first steppingstone in part of a much larger overall oper-ation, an operation first conceived in North Africa in the first weeks of 1943.

Held at the Anfa Hotel in Casablanca, French Morocco, the Casablanca Conference began on January 14, 1943. The gathering's purpose was to plan the Allied European strategy, as the United States had finally seized momentum and officially entered the offensive phase of the war. President Franklin Roosevelt, British Prime Minister Winston Churchill, and the U.S. Combined Chiefs of Staff were among those in attendance. According to Samuel Eliot Morison's history, at this critical hour long-term plans for victory needed to be drawn up for both theaters of war with two essential questions in mind: "Where do we go from here?" and "How are we going to do it?" On January 23, 1943, the Combined Chiefs of Staff issued a report that Roosevelt and Churchill both approved. The first step of that report was an operation "to continue 'Operation Watchtower' up from Guadalcanal and New Guinea until Rabaul was taken and the Bismarcks Barrier broken."[21]

Later codenamed "Operation Cartwheel," Cartwheel was the overall island-hopping campaign meant to surround, neutralize, and dismantle the Japanese airbase on Rabaul, New Britain, located about six hundred miles northwest of Guadalcanal. MacArthur himself referred to the strategy behind the campaign as "leapfrogging," and as such stated, "a concept as ancient as war itself . . . the process of transferring troops by sea as well as by land appeared to conceal the fact that the system was merely that of envelopment applied to a new type of battle area. It has always proved the ideal method for success by inferior but faster-moving forces."[22]

After it fell, Guadalcanal became a critical supply and training base for Allied operations in the Pacific. Guadalcanal's neighbor, Florida Island, located several miles across the Ironbottom Sound, was also transformed into a major center for supplies, fuel, fresh water, equipment, spare parts, and materials needed to repair ships. Just as important, the miniature island of Tulagi within its bay was made into the first forward operating base for patrol torpedo boats (PT boats) in the South Pacific. As the Allies pushed northwest during Operation Cartwheel, the LCIs were tasked with the important job of resupplying at Florida Island and delivering those supplies to the front lines. This was of course when the LCIs were not engaged in combat themselves.

Even if the tired LCIs weren't fighting, they were still desperately needed for their back-and-forth labor. Stephen would spend the next year and a half in the fight to achieve victory in Operation Cartwheel.

———

In the weeks following the Casablanca Conference, just as the infantrymen and Marines were wrapping up the fighting on Guadalcanal, Stephen received his very first authorized allowance of $2.75 per day—officially granted on February 5, 1943. If in his exact situation today (2022), adjusting for inflation, that $2.75 would be the equivalent of earning an allowance of about $42 a day—less than minimum wage. Not exactly the fair compensation one would imagine was deserved and appropriate for those sailors willing to sacrifice their lives to fight Hitler, Mussolini, and the Imperial Japanese.

Days before they departed Galveston, Stephen and his friends decided to get tattoos. According to LCI 70 veteran Royal Wetzel, "Everybody said you have to have a tattoo before you went overseas."[23] Stephen's female artist that night was covered in tattoos from head to toe. Considering the times, he remembered she even had tattoos on her face. He ended up getting a tattoo of a bald eagle perched on a tree branch on his right upper arm. Branded below the eagle's talons was the word "PEACE." He got a second tattoo on his left upper arm of a dagger.

"So, from (Galveston), we went down to Panama Locks. And we crossed the locks. It took about a week to get through," Stephen said. However, the LCI 329 experienced a major issue in their trip through the Panama Canal that delayed their trip to the South Pacific. It did not, however, slow down the rest of the ships in LCI Flotilla Five. On February 25, 1943, the forty-four-ship convoy that made up Flotilla Five had proceeded out of the Gulf of Panama. The LCI (L) 334 acted as division leader of the column on the starboard side of the LCI (L) 328, which acted as the LCI group guide. For the next ten days, the column of LCIs steamed toward Bora Bora without the LCI 329. Stephen and the crew were on their own.

Stephen's buddy from San Francisco, Elmo Angelo Pucci, explained in William L. McGee's book years later why their LCI was not part of the convoy that left Panama. "We got stuck at the Panama Canal. We got caught in some wind there and fouled up a screw. Had to go back to Coco Solo and go into dry-dock. Next thing you know, we were all by ourselves. All the other LCIs had continued on. We went the rest of the way to New Caledonia all by ourselves cruising at 12 to 14 knots tops."[24]

Pucci, a handsome young Italian man, was the ship's cook while aboard the LCI 329. He attended cooks and bakers school while in San Diego, but said it hardly prepared one for life as a cook because "When you went to Cooks and Bakers School, you didn't know anything about cooking. Maybe you made coffee in the galley or something." He added, "You didn't even learn how to cut a piece of meat. Of course, I had a little knowledge of that because I worked in the meat industry before I went into the Service."[25] Upon first transferring aboard the 329, Pucci quickly realized the situation: "There was nobody else. You're the only cook. So you gotta be the cook. You gotta be the baker. You gotta be the commissary steward and storekeeper. The whole kaboodle. It was a tough job. You went to work seven days a week." Not to mention the hours were grueling. He remembered, "Get up 5:00 in the morning. Cook breakfast, cook lunch, cook dinner. In between you had to bake bread. You had to learn that on your own. I learned all that stuff on my own." Pucci recalled the laborious task of feeding all those men every single day, saying, "We had an extra officer that made 26 (crewmen). I had no assistant cook. I was lucky I had a mess cook who didn't know how to wash dishes. I had to wash dishes. They were afraid to put their hands in the hot water."[26]

LCI 448 veteran Chuck Savard remembered the cooks trying to cook aboard his ship during a typhoon. Savard wrote, "I was in the galley when we rolled over so far that the galley port scooped up water from the huge waves and flooded the galley sink."[27] On top of adjusting to the turbulence and instability of preparing food while being constantly thrown about the sea, the cooks also had to learn how to wheedle delicacies from other ships. Constantly short on even the bare necessities, Pucci said, "We didn't have sea rations. We used to have raw meat. I used to throw all kind of supplies, anything I could get ahold of." When asked if the larger American ships helped them out, Pucci responded, "Very seldom. Sometimes when we got a chance, we got some fresh fruit . . . [cooks] swiped whatever they could swipe." William H. McCracken, a lieutenant (jg) on the LCI (R) 1030, agreed with Pucci's assessment writing:

> LCI's in World War II were unfortunately doomed to bear the cross of procuring, oftimes begging, survival supplies and nourishment from any nearby larger naval vessel . . . LCI survival was frequently supported by negotiated barter exchanges among sister ships, involving unexplainable excess ten year inventories of say frozen chicken feet and rustproofing paint that none wanted to put on anyway.[28]

Royal Wetzel, a Ship's Cook 3/c aboard the LCI 70—who humorously enough bears a striking resemblance to the famous cartoon character Popeye—also shared Pucci's struggle. In an interview conducted in 2011, Wetzel said, "The food wasn't too good though. I was the cook, and I didn't have anything to cook with. We had what they called mutton. That's what we had. I would go fill the guys' trays up for them and they would throw it over the side." He also remembered that since they were on such a small ship "we didn't have any place to eat. Wherever you could find a place to eat on, you would sit in the ready boxes and eat."[29] After the war, many LCI veterans swore they never again ate mutton.

Pucci added, "I was so glad to get off that [LCI] in 15 months." Later in the war, Pucci would transfer off the 329—and LCIs all together—eventually becoming part of the LCT(5) Flotilla Staff and LCT 182. Nonetheless, even though Pucci had to experience all those singular miseries of being the only cook aboard his LCI, he continuously found a way to make it work. "I always had supplies," he remembered. "In fact when we were down there in New Georgia up in the slough, I had food to last us 30 days. Some of the boats ran out. They were eating beans. They come over to me beggin' for sugar and stuff."[30]

But Pucci could always be counted on to look out for Stephen and his crewmates. "I had stores stacked all over that LCI", he assured. "Wherever I could grab them, I grabbed them." But make no mistake, "It was one of the toughest jobs that I've ever had in my life," he recalled.[31]

———

To understand the duties of the men aboard the LCI 329, one must understand the layout and design of the LCI 1-350-class LCI. The LCI 329, like all LCIs, had three levels that were connected by a series of stairs and ladders. The lowest level of the ship, or first level, was where the crew's quarters, troops' quarters, and engine room were located. This first level was located belowdecks and required a ladder to get down to it. The crew's quarters, where the enlisted men slept, were filled with bunks stacked vertically, in tiers four-high, with very little space in between. According to veteran George Weber, "You could not raise your knees in turning over without jabbing the sailor above you, and you got jabbed by the guy below you. Hefty crew members had difficulty getting in and out of their bunks." Since the engine room was right next to the crew's quarters, it often stunk of diesel oil and sulfuric acid from charging batteries. Weber added, "These fumes were added to the other accumulated odors filling the crowded compartment: foods, rank body odors (no showers, other than saltwater ones), and the flatulence produced by the traditional naval Wednesday and Saturday breakfasts of baked beans, cornbread and coffee. The ventilation system—never outstanding—was sorely inadequate for handling this combination of smells."[32]

Above the first level was the main deck, or second level. Upon climbing up the ladder to reach the main deck, this second level was most often crowded and cluttered with equipment, supplies, and ammunition. This was the level where most of the activity occurred abovedeck. This was also the area where the crew would relax during breaks, whether it was at the forward part of the ship between the deckhouse and the bow, known as the *well deck*, or between the deck-house and the rear of the ship, known as the *fantail*.

LANDING CRAFT, INFANTRY (LARGE) **LCI (L) 1–350**

A blueprint of the LCI (L) 1-350-class design. LCI 329 was among these first LCIs manufactured which had a square conning tower. A round conning tower design later replaced the square design, beginning with LCI 351 and continued through LCI 1098. (*Allied Landing Craft of World War Two*, U.S. Naval Institute Press)

The deckhouse, an enclosed area in the middle of the ship directly behind the conning tower, contained the wardroom, radio room, officers' quarters, crew's bathroom, and galley. The galley was where the ship's cooks like Pucci prepared breakfast, lunch, and dinner. At mealtimes, the crew would line up and the cooks would fill their metal trays with "slop" as they shuffled past. As Royal Wetzel mentioned, there was hardly anywhere around the mess area for the sailors to eat before they'd make their way back to wherever they were needed. The men would sit wherever they could find an open spot. The second level was also where the 20mm and .50 caliber machine guns were mounted, including Stephen's 20mm gun that was positioned on the gun deck (atop the deckhouse). Each of the four 20mm guns rested in an enclosed mount, or gun tub, which was essentially just a thin layer of plastic armor surrounding and protecting the guns and four men operating each one. These powerful and versatile 20mm guns could be used to attack both air and surface targets.

Rising from the gundeck at the middle of the ship was the conning tower—a ten-foot-high square-shaped structure speckled with seven porthole windows (three in front and two on each side). Within the conning tower was the pilothouse, a cramped, enclosed room containing the ship's navigational equipment and magnetic compass. The pilothouse contained the levers for the helmsman (most often a quartermaster) to control and steer the ship. At the very top of the LCI 329's square conning tower was the third and highest level, the bridge. Ensign Read Dunn, Jr., the executive officer of the LCI 335 stated in his autobiography, "There was an open bridge above the conning tower from which the ship could also be controlled in good weather or in restricted waters. The ship was steered by twin rudders controlled by a solenoid switch in the [pilothouse] which was operated by turning a handle. It had no steering wheel."[33] The bridge was the location where the commanding officers, signalman, and often chief quartermaster stood during battles and was considered the brain of the ship. There, officers would give orders to the quartermaster steering the ship in the pilothouse below. Those atop the bridge were able to communicate with the pilothouse, engine room, and all other parts of the ship using a sound-powered telephone. They could also communicate with other ships using semaphore flags or the signal light that was mounted on the edge of the bridge. While the LCI 329 was engaged in combat, officers on the bridge and on every deck wore headsets with microphones so that they could communicate with each other.

And always above the crew was Old Glory. As Pulitzer Prize winning author Mitch Weiss once described it, "Towering over the [LCI] was the mast. (It was welded to the gun deck and attached to the side of the conning tower.) Like a clothesline with laundry, the tall pole was crisscrossed by a series of lanyards to hold up flags of various sizes, shapes, and colors to send messages between ships. At the very top of the mast: the U.S. flag."[34]

Name	Rating/Rank	Date of Enlistment	Date First Received Aboard	Place of Enlistment
ILLING, William Arthur, Commanding Officer	Lieutenant (junior grade)	12/12/1941	11/8/1942	Little Rock, AR
LOVE Jr., Frank G., Executive Officer	Ensign	7/1/1942	11/8/1942	Dallas, TX
POST, Albert J., Engineering Officer	Ensign (Machinist)	9/21/1933	6/2/1943	Bremerton, WA
ANDERSON, Helmer R.	Electrician's Mate, 3rd Class	3/14/1942	11/8/1942	Des Moines, IA
ANDERSON, Rex W.	Seaman, 2nd Class	3/23/1942	11/8/1942	Denver, CO
BIHLMAN, William F.	Apprentice Seaman	8/26/1942	11/8/1942	Cincinnati, OH
COLBERT, Joseph C.	Seaman, 2nd Class	2/2/1942	11/8/1942	Indianapolis, IN
CRANGLE, Joseph M.	Seaman, 2nd Class	1/30/1942	11/8/1942	Des Moines, IA
DOERR, Charles E.	Fireman, 2nd Class	6/2/1942	11/8/1942	Springfield, IL
FEENEY, Luke J.	Seaman, 2nd Class	4/9/1942	11/8/1942	New York, NY
FEHRING, Leonard H.	Fireman, 2nd Class	6/16/1942	11/8/1942	Oklahoma City, OK
GANZBERGER, Stephen	Apprentice Seaman	8/26/1942	11/8/1942	Detroit, MI
HAMPSON, Richard P.	Motor Machinist's Mate, 2nd Class	6/15/1942	11/8/1942	Cincinnati, OH
HOLLAND, Bruce S.	Fireman, 1st Class	6/11/1942	11/8/1942	Dallas, TX
LANDRY, Harold E.	Fireman, 2nd Class	6/14/1942	11/8/1942	New Orleans, LA
LEMASTUS, James W.	Apprentice Seaman	8/27/1942	11/8/1942	Louisville, KY
PACK, Woodrow Wilson	Apprentice Seaman	8/27/1942	11/8/1942	Toledo, OH
POWELL, John W.	Apprentice Seaman	8/26/1942	11/8/1942	Detroit, MI
PUCCI, Elmo Angelo	Seaman, 2nd Class	4/15/1942	11/8/1942	San Francisco, CA
RANDALL, Harold C.	Fireman, 2nd Class	6/5/1942	11/8/1942	Lima, OH
SLEIGHT, James R.	Signalman, 3rd Class	6/13/1942	11/8/1942	Baltimore, MD
STANLEY, Val G.	Fireman, 3rd Class	6/2/1942	11/8/1942	Marion, IL
WILSON, Charles R.	Seaman, 2nd Class	4/2/1942	11/21/1942	San Diego, CA

LCI (L) 329 ORIGINAL CREW LIST, PIER 18, GALVESTON, TEXAS. *Note*: Quartermaster 2/c Floyd H. Mirick and Radioman 3/c Stanley DeJewski are not included. DeJewski reported aboard February 25, 1943, and Mirick reported aboard on March 9, 1943, in the Panama Canal Zone. LCI (L) 329 was built at Brown Shipbuilding Co. in Houston, Texas, and commissioned on November 8, 1942. The men highlighted remained aboard the LCI (L) 329 for the entire year of 1943. Dates are in month/day/year format. The crew's information comes from the LCI 329's November 1942 muster roll.

Crew Formations (April 1943)

The crew of the LCI 329, like every ship in the Navy, was organized by a strict hierarchy of command. The officers were in charge at the top of the hierarchy with the captain, or skipper, sitting at the top. The 329's skipper, Lieutenant (jg) William Arthur Illing, handled navigation and communication duties on the bridge atop the conning tower. If General Quarters was suddenly sounded, the skipper would relieve the officer of the deck on the bridge. The second in command, Lieutenant (jg) Frank G. Love Jr., the Executive Officer (XO), was from Texas and a graduate of Southern Methodist University. Love's duties included censoring the crew's mail and supervising the "deck gang," which mainly consisted of the seamen and gunner's mates. The Executive Officer led gunnery training drills but also handled disciplining the sailors who ran into trouble. When in combat, Love spotted for the numbers one and two 20mm guns. Third in command, Ensign Albert J. Post, served as the engineering officer. Post supervised the sailors who operated the engines belowdecks in the engine room, but during combat would spot for the numbers three and four 20mm guns. As the LCIs only had three officers, the Engineering Officer was typically also the Commissary Officer that handled supplies. Most officers assigned to LCIs benefited from an abbreviated three-month wartime program involving universities around the nation.[35] Called "90-day wonders," they earned a commission, and oftentimes even the command of a ship, within a rather quick ninety days—and were usually totally green. Typically, each junior officer would follow a rotation, serving four hours on watch and eight hours off.

All three officers of the 329 worked closely with the noncommissioned officers, or petty officers. Typically, there were five of these petty officers aboard an LCI, composed of the highest ranked enlisted men. The petty officers most commonly consisted of one signalman, one electrician's mate, one radioman, one coxswain or boatswain's mate, and one motor machinist's mate. If an LCI was assigned a yeoman, he was also commonly chosen to be a petty officer. The petty officers carried out the day-to-day supervision of the "deck gang," such as cleaning, painting, training, and conducting drills. According to author Mitch Wiess, the petty officers also ensured that every man "knew how to use the vital items like fire extinguishers, life rafts, and storage lockers."[36] Those responsibilities fell on the 329's two Coxswains, Joseph C. Colbert and Allan E. Johnson.

The bulk of the crew was made up of the common enlisted men —the sailors fresh out of boot camp like Stephen. On the 329, there were twenty-two of these so-called blue jackets. The enlisted men standing at the highest level of the ship were the signalmen, or as they were jokingly nicknamed, "skivvy waivers," as their semaphore flags resembled underwear. The signalmen were needed atop the bridge of the conning tower at the highest point of the ship because they were in charge of communicating with other ships in the fleet using either a series of semaphore flags or by sending Morse code with a signal light that was mounted to the bridge. A signalman sent messages using the semaphore flags by positioning two flags in places in relation to his body. Each position the flag was in represented one of the letters of the alphabet. In instances of decreased visibility like nighttime, a signalman could also use the round signal lamp to send messages by opening and closing the shutters using handles on both sides. They'd open the shutters quickly for a dot and slowly for a dash. The alphabet consisted of a series of dots and dashes using Morse code.[37] The two signalmen aboard the LCI 329 were Signalman 2/c Rex W. Anderson and Signalman 3/c Charles R. Wilson. While the ship was underway, there needed to be at least one signalman on duty with the officer atop the bridge, and on occasion, a quartermaster joined them.

Directly below the officers and signalmen was the quartermaster within the pilothouse—the glassed-in space below the conning tower's bridge. As the officer barked orders down through a tube, the quartermaster listening below handled the navigation and steering, but also updated the charts and helped with the visual communication. According to author Louis Harlan, "The quartermaster adjusted the ship's rudder or steered on a compass point according to the captain's order. He transmitted the captain's order as to speed down to the engine room by means of the engine-order telegraph, a circular gadget that rang a bell and set the engine speed anywhere from all ahead flank to all back full." In his book, *All at Sea*, Harlan argued that sometimes drastic and unorthodox variations in the LCI's maneuverability were necessary because "unlike a car, a ship had no brakes."[38] On March 9, 1943, Quartermaster 2/c Floyd H. Mirick from Pineville, Kentucky reported aboard the 329 as its newest crewmember. Stephen would spend the year of 1943 striking (training) with Mirick to become a quartermaster.

Several different groups of men operated on the main level amidship within the deckhouse. In the galley, were the cooks, such as Stephen's buddy, Ship's Cook 3/c Elmo Pucci. The deckhouse also contained the radio room, or radio shack, where the radiomen sat wearing headphones in front of the short- and long-range radios used to send and receive messages. If the crew wanted the latest current events in the war, they could usually turn to Radioman 3/c Paul D. Driscoll, and later Radioman 3/c Stanley W. DeJewski to deliver the latest rumors or "gator gossip," since those men were constantly listening to Navy broadcasts. At times, the crew used their radio to tune in to Tokyo Rose, Japan's voice of propaganda used to taunt and frighten American forces during the war through her radio broadcasts.

Deep belowdecks, at the lowest level of the ship, resided the loud, grimy, sweltering engine room, where the speed and electricity of the ship was manually controlled. Within the engine room were the motor macs—known as the "Black Gang"—who labored away in the darkness operating the diesel-powered engines. Their uniform of the day was usually cut-off dungarees and wooden sandals. Profusely sweating and always covered in engine soot, Motor Machinist's Mate 2/c Leonard H. Fehring, Motor Machinist's Mate 2/c Bruce S. Holland, and ranking Motor Machinist's Mate 1/c Richard P. Hampson always seemed to be thirsty. As motor macs, they had perhaps the most miserable and undesirable job of all. If the ship were to sink, those in the engine room would be in the most vulnerable and perilous spot of all, as they would be trapped belowdecks. Assisting the black gang within the 329's noisy engine room was Electrician's Mate 3/c Helmer "H.R." Anderson who stood before the massive electrical switch-board. Anderson managed the equipment and generators that provided power to every part of the ship. The designer of the LCI, John C. Niedermair, once said in his oral history in 1975, "I was told later that anybody who had duty in that engine room couldn't hear for the longest time [afterwards]."[39]

Spread throughout the rest of the ship were the remaining seven seamen and three firemen that made up the deck gang, which as mentioned were responsible for the daily duties of the ship. These men were usually training for a specific job like gunner's mate. The trio of firemen aboard the 329 consisted of Fireman 2/c Charles E. Doerr, Fireman 1/c Harold E. Landry, and Fireman 1/c Harold C.

Randall. The seven seamen rounding out the deck gang were Seaman 2/c Stephen Ganzberger, William F. Bihlman, Joseph M. Crangle, James W. Lemastus, Val G. Stanley, Seaman 1/c Luke J. Feeney, and Woodrow Wilson Pack.

And always on the mind of each individual crewman, day after day, was the ship's "scuttlebutt," or latest rumors. Regardless of a man's rating or his place in the hierarchy of command, he soon learned that a ship is not a Navy ship without scuttlebutt, banter, reminiscences and, above all, rich and colorful griping. Or as correspondent Ernie Pyle put it, "I don't know how we would endure war without its rumors."[40]

———

One of the requirements of the Navy is that all commissioned ships and landing craft must maintain an official account of the day's events, known as a deck log, which typically summarizes the ship's location, duties, movements, and happenings in four-hour increments. As the war intensified in the years to come, the deck logs of Stephen's LCIs would at times be detailed by the minute. However, the first official record of the deck log of the LCI (L) 329 simply began on March 16, 1943, with a single entry:

En route through Panama Canal

Followed by the next day's entry:

En route to Pacific

And with that their long Pacific journey began. The LCI 329 needed to be accompanied in its journey by USS *Sepulga*, an oiler ship, because the LCI constantly required refueling as it sailed across the vast Pacific. On March 22, 1943, at 104 degrees west longitude, Stephen and his shipmates crossed the equator and were duly initiated as what's known as a "shellback," a Navy term used to describe a person who has crossed the equator for the first time. Though not touched upon in this book, the initiation ceremony for these men who'd never crossed the equator before ("pollywogs") is known to be quite bizarre.

It was a long and uncomfortable trip across the Pacific in that tiny LCI. The crew of twenty-two men and three officers felt every wave due to its flat bottom. Any significant movement in the water would cause the ship to toss about. LCIs didn't have the rolling motion of a destroyer. The flat bottoms on LCIs were not displacement hulls like on typical Navy ships. LCIs rode on top of the water like wood chips. Flat-bottomed landing craft did not cut smoothly through the water. Instead they belly-flopped their way up and down, as if trying to beat each wave to death. As Louis Harlan remembered, "'The sea was so rough it was like riding over a cobblestone road in a one-wheeled wagon.'" LCI veterans would unanimously agree years later that the word "rough" didn't even begin to describe the journey across the vast ocean. And when the LCIs travelled at night, the ships were completely darkened to protect them from being seen by the enemy. As an additional safety measure, all porthole windows were also covered with blackout shades. Once "lights out" was ordered not even a match could be struck topside to light a cigarette.[41]

And then there was the seasickness. For the men who were "not seasoned mariners" as Commodore John H. Morrill described them, "seasickness feels like they have just spewed out their entire stomach." He wrote, "A few feel that they have lost the desire to live. For all of them, any kind of work is next to impossible." Morrill added, "For those aboard ship who were not seasick, it meant a sleepless night."[42] Vice Admiral Lorenzo Sabin Jr., commander of LCI Flotilla Two, spoke bluntly of his crew's seasickness aboard his ship in a letter to his friend, Commander D.C. Varion, saying, "Did you ever hear about sailors with all the courage in the world but no guts? If you haven't, you will, because my sailors lost all their guts twenty-four hours after sailing.... You know of course, how interestingly stuffy, cramped, and uncomfortable [LCIs] are." Sabin remembered a majority "were so seasick most of them couldn't get out of their bunks."[43] Other men simply carried around buckets to throw up in.

Like Morrill and Sabin, Stephen witnessed the seasickness, too. While transporting a group of twelve Marines to New Caledonia, Stephen recalled, "I really felt sorry for this one, 'cause he was sick all the way down to Noumea, New Caledonia. [He was] throwing up blood at the end. That's how sick he was." On the LCI 555's journey to Europe, Louis Harlan wrote, "Practically every man aboard was sick. Smells of vomit and fuel oil smote the nostrils in a deadly combination." To make matters worse, Harlan said his crew "might have expected breakfast, but the cook did not stir from his bunk. He was among the sickest of all. Kitchen odors only caused him to revisit his seasickness." Having an entirely seasick crew caused quite an issue according to Harlan because "There had been no hot meals for three days, and the galley smells were almost unbearable. Clutter was everywhere, and there was no will to clean it up."[44] A war correspondent aboard the LCI 226 wrote that much of the crew was unashamedly sick. He remembered a crewmember who got so thoroughly seasick he lost his set of false teeth to the sea, which the crewman was unable to replace until his ship reached Australia.[45] Unlucky LCI 750 veteran Bob Petit, at the time a young boy who managed to enlist in the Navy at only thirteen years old, wrote years later that while crossing the Pacific aboard the oiler USS *Tallulah*, "I started getting seasick almost immediately. Of course, I didn't know you were supposed to go to the lee side of the ship to throw up. I went to the wind-ward side, and it all blew back in my face."[46]

And then there was the issue with the ship's bathroom, or crew's head. Simply put, according to author Mitch Weiss, "The bathroom was a hellhole." He described how the men on the LCI 449 "shared two showers, four washbasins, a urinal trough, and six toilets, nothing more than a slat of wood with holes. Seawater ran continually beneath and flushed the waste over the edge. The ship could distill three thousand gallons of freshwater each day, but it was strictly rationed, mostly for drinking and cooking, because the distilling equipment broke down often, and no one wanted to be stuck on the ocean without freshwater."[47] Only thirty-seven tons of fresh water could be carried by an LCI at a time. "Not much water for a trip across the Pacific," added LCI 74 veteran Robert Kirsch. However, life aboard Stephen's more primitive LCI (L) 1-350-class ship had less space than the LCI 449, which meant the bathroom was even more cramped. Kirsch, who sailed onboard a ship with the same LCI design as Stephen's LCI 329, explained:

> The head consisted of one shower stall, one washbasin with a mirror, a hog trough with 3 seats which consisted of two barrel staves across the trough for you to sit on. . . . Flushing was simple, a pipe was hooked to one end of the trough and exited out the other side of the ship. The fire main water continually flushed the trough with salt water. Now mind you, we had a crew of 24 enlisted men using this very small head.[48]

Aboard LCI 555, "Showers were out of the question, not only because we had to conserve fresh water," Louis Harlan remembered, "but because we would have bounced from wall to wall in the seawater shower stalls."[49] And if an LCI's distilling equipment was broken or missing, which happened often, the men were forced to bathe in seawater using the occasional torrential downpour to rinse off the salt. On the LCI 628, Thomas Woodstrup said, "We could take a salt water shower anytime and a fresh water shower every third day when under way. Our head was two troughs with a wooden slat for each cheek."[50] After a two-week stretch of no land in sight, "Big Bill" Athan of the LCI 446 hilariously recalled, "After 2 weeks of salt water showers, [and] water rationing, we docked at Bora Bora, Society Islands. At first I thought we had died and gone to heaven! We were given fresh fruits, milk and ice cream. We had a diarrhea party next day with only 6 hoppers! What a time we had."[51] All the while, any man struggling to shower or relieve himself would be in Sabin's words, "Bounding and pounding: twisting and twirling; rolling, buckling, . . . [d]ay after day; night after night; week after week. Tossing, turning and twisting. Pitching, pounding and rolling. Up by the bow, down by the stern."[52]

Against the fury of the smashing waves, the LCI 329 made its way into the port of Bora Bora, Society Islands (today Polynesia) on April 5, 1943, where the crew spent the next week repairing, restocking, and refueling—which included taking aboard 17,000 gallons of diesel fuel. Robert Rosenwald of the LCI (L) 1008, like Stephen and Big Bill, also traveled from Panama to Bora Bora. As he crossed the equator heading for Bora Bora, Rosenwald remembered, "I would watch the Big Dipper constellation becoming lower and lower each night until it no longer rose above the horizon. Then the Southern Cross started to show above the horizon. From that time on . . . the Southern Cross was there."[53]

Stephen and the 329 departed Bora Bora on April 12 and arrived in Pago Pago, Samoa, four days later on the 16th. It was here in Pago Pago—the 329's second stop—that Stephen remembered most vividly of all. "Most beautiful place I've ever seen," he recalled with clarity on several occasions. "It was so pretty over there. You'd see a big mountain, and it was all a big harbor, it was all nice white beach. It was such a pretty place." He even recalled a run-in with a shark there. "That's where I seen a shark. It was about twelve feet long. One time, we used to go on what they called garbage detail, and I'd get up on the gunwale (far rear of ship), and two of us would take the garbage and dump it over the side. And the moment I dumped it over the side, I looked down there, and there's that big [shark] just waiting for the garbage. It was like a Mako Shark, it was a big one."

Two days after arriving in Pago Pago, the 329 took on twelve Marines, including the violently seasick one, for transportation to New Caledonia on April 18.

"We only stayed a week," and on April 21, Stephen recalled, "We pulled up anchor and went to Fiji Islands." On April 24, en route to New Caledonia, the LCI 329 made a pit stop at Suva, Fiji—about seven hundred fifty miles to the southwest of Pago Pago, Samoa. They arrived in Fiji shortly after 1:30 p.m. The crew would spend the next day, Easter Sunday, moored in Kings Wharf, just off the coast of Suva,

Fiji. "We stayed maybe three or four days," Stephen remembered of their short stay in Fiji. "They filled us full of water and they filled our fuel up, and we picked up supplies."

———

As the crew of the LCI 329 awoke on the morning of April 26, 1943, a crucial decision had been made in Brisbane by two top military leaders who had just met. The outcome of this meeting ultimately determined the direction of all forces that were in the Pacific, including the LCI 329. Vice Admiral William F. Halsey, Commander of the South Pacific Area and Third Fleet, and General Douglas MacArthur, Commander of the Southwest Pacific Area and Seventh Fleet, had completed their discussions and finalized their plan on how to surround and strangle the Japanese stronghold base of Rabaul, New Britain—essentially cutting it off from supplies and reinforcements. An island-hopping strategy was devised. When the two legendary men parted ways, it officially marked the beginning of the *Elkton III* plan. *Elkton III* became "Operation Cartwheel," a two-pronged, simultaneous attack to seize enemy-held northeast New Guinea (Southwest Pacific Area) and the remaining Solomon Islands (South Pacific Area). MacArthur's forces would drive up New Guinea as Halsey's forces drove up the Solomon Islands. Their goal was clear: strangle Rabaul. Neutralize it. Surround and encircle it, without directly invading it. Use combined Allied strength and superior forces over a vast area to seize strategically important but weakly defended areas, while bypassing heavily fortified positions. In 1958, fifteen years later, Major General R. W. Stephens, Chief of Military History, pointed out that a vital component to the Cartwheel strategy was not only invading the right places, but also avoiding certain others. He wrote that Cartwheel "established the technique of bypassing [Japanese] strongholds, including finally Rabaul itself, and threw [Japan] on the defensive."[54] Or as General MacArthur plainly put it, "hit 'em where they ain't." In his 1964 book, *Reminiscences*, MacArthur explained, "Each phase of advance had as its objective an airfield which could serve as a steppingstone to the next advance. . . . [A]s this air line moved forward, naval forces under newly established air cover began to regain the sea lanes, which had been the undisputed arteries of the enemy's far-flung positions."[55] This strategy would cut the Japanese off from their supply lines and leave them isolated and vulnerable with no other option but to remain entrenched where they were. Stephen would be involved in this strangling of Rabaul at each step along the way, even patrolling the channel off Rabaul itself in just over a year's time.

As Halsey's and MacArthur's forces converged toward each other, the next year would be a long and bloody one. Already men and supplies were moving into the forward areas.

Stephen and the LCI (L) 329 departed Fiji on April 28 for New Caledonia. On May 2, Stephen's LCI anchored at Port Noumea, New Caledonia, in three fathoms of water. The transportation of the twelve marines that Stephen's LCI had picked up in Pago Pago, including the violently seasick Marine who had vomited blood, was completed successfully. After spending two weeks together with the Marines—making an already cramped ship even more unthinkably cramped—Stephen bid them farewell. He would never see them again.

As the winds of war carried LCI sailors across two vast oceans into the unexpected, for nearly all of them, including Stephen, this was the men's first time away from home. Though they'd never outwardly show it for fear of appearing soft or weak, homesickness inevitably found its way to each one. Louis Harlan may have best expressed the average LCI sailor's state of mind during this time. On their first trip across the Atlantic Ocean aboard the LCI 555, Harlan remembered, "As I stood watch there under the everlasting stars, I was conscious of having left behind everything that was familiar. I was on unknown waters, headed for an unknown land."[56]

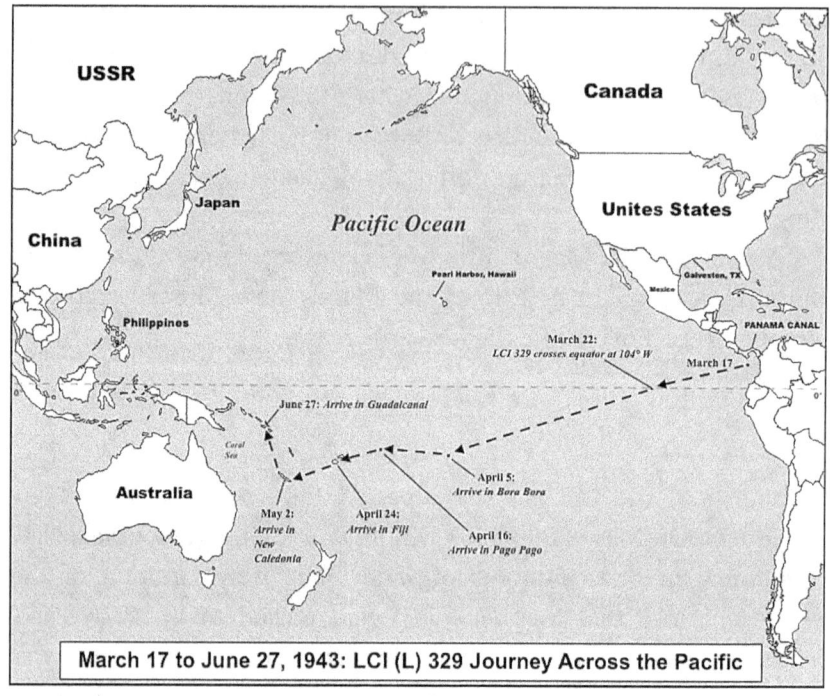

Modified version of original map: https://d-maps.com/carte.php?num_car=
3260&lang=en

3

JULY 1943: THE NEW GEORGIA CAMPAIGN

"The price of Rendova was the blood and sweat of heroes."

– Robert P. Patterson, Under Secretary of War (1943)

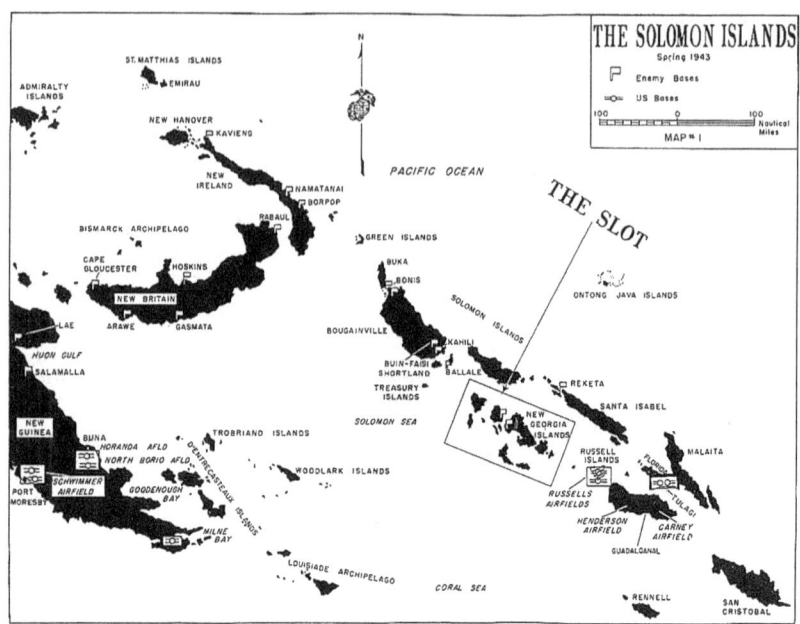

A map of the Solomon Islands, part of the South Pacific Area of Operations, showing the situation in Spring 1943 before the Allied invasion of the highlighted New Georgia Island group. (U.S. Marine Corps photo, 1952)

Stephen and the crew of the LCI 329 spent the next month and a half at New Caledonia training in their various duties—everything from radio school to gunnery school. The men sharpened their skills needed for the impending action they would soon see against the Japanese. Now in New Caledonia, the crew of the 329 was moving ever closer to the war raging in the Solomon Islands.

"We'd still get air raids while [I] was there . . . because you were getting pretty close to the front lines," Stephen recalled.

On June 23, 1943, the LCI 329 received orders to depart New Caledonia. This was what the men had been waiting for. It was finally time for them to join the fight against the Japanese. The 329 soon joined LCIs 328, 66, and 331 (Section I) and LCIs 335, 24, 22, and 21 (Section II). The two columns of LCIs were bound for the Solomon

Islands as part of Task Unit 32.5.1 under the command of their beloved Lieutenant Commander Alfred Vernon Jannotta.

Four days later, Stephen and his buddies arrived at Lunga Point, Guadalcanal, on the morning of June 27, where the American forces had been waiting for the 329 and the other LCIs. The area was buzzing with activity. The Allies had just begun the seizure of their next major target in the South Pacific—the New Georgia Islands. Many squadrons of planes could be seen overhead. Unknown to Stephen and the crew of the 329, six days earlier the 4th Marine Raider Battalion struck Segi Point on New Georgia proper, the largest of the New Georgia Islands, in pursuit of Viru Harbor. The LCI 329 had arrived just in time for the opening days of the New Georgia campaign—dubbed "Operation Toenails."

The New Georgia campaign was an extraordinarily special landmark for the Amphibious Forces. It was the first campaign in the Pacific in which the Allies utilized LCIs to land troops on enemy beaches. It was the LCIs' big debut. Any engagements with the Japanese would be among the first combat seen by LCIs in the Pacific Theater.[1]

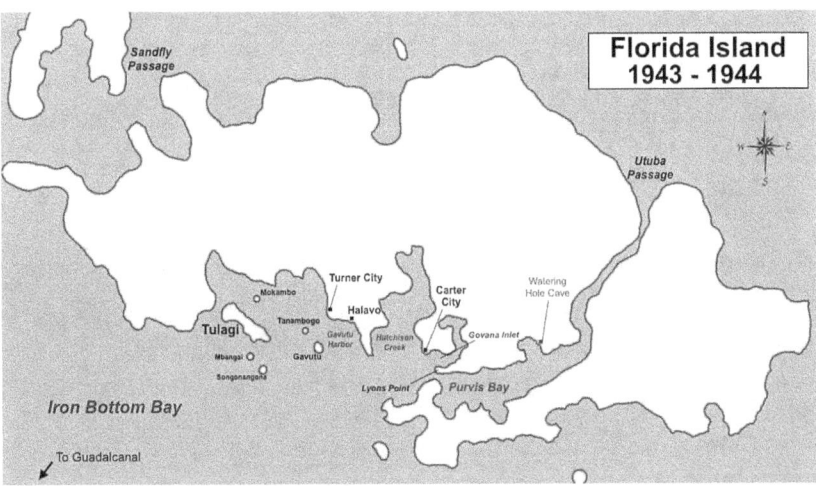

Across from Guadalcanal sits another island named Florida Island, today known as Nggela Island. Florida Island forms a sort of semi-circle around its entrance bay where an even tinier island is nestled within, known as Tulagi Island. Tulagi was the first island ships would see on their port side upon arriving. As the LCI 329 entered Purvis Bay from Ironbottom Sound on June 29 to take on fresh water, Stephen looked up at the entrance of Tulagi. What he saw next would be seared into his memory for the rest of his life and would become one of his most frequently referenced sea stories from the war. A large sign was posted on Tulagi for all men to see as they entered, a stark reminder for why they were all there. In big bold letters, the sign read, "ADMIRAL HALSEY says 'kill japs, kill japs, KILL MORE JAPS'! You will help kill the yellow bastards if you do your job well." Future U.S. President John F. Kennedy, then a lieutenant and skipper of a PT boat, who was also in the South Pacific at that time, recalled years later the first time he saw the sign, remarking, "It went right through you."[2]

The sign posted outside Admiral William "Bull" Halsey's headquarters on Tulagi, Florida Island, circa 1943 (cropped). (U.S. National Archives photo No. 80-G-259446)

Fleet Admiral William Frederick "Bull" Halsey Jr., circa 1945.
(U.S. National Archives photo No. 80-G-K-15137)

After the LCI 329 had resupplied at Florida Island, they departed for the Russell Islands. They arrived in Maquitti Bay, Pavuvu, Russell Islands on July 3. They loaded up some of the last remnants of the 169th Field Artillery Battalion, part of the 43rd Infantry Division. The 43rd Division had spent the last four days clearing Rendova Island of Japanese.

Historians have written much about the five initial landings in the New Georgia Islands, which occurred at Wickham Anchorage, Viru Harbor, Segi Point, Roviana Lagoon, and Rendova—but they all tend to agree on one thing: the Rendova landing actually paid off.

Rendova is a tiny island that lies just to the south of the main island of New Georgia proper with only five miles of channel separating the two islands from each other. Despite its rather meager size, the Allies needed to capture Rendova at all costs. There was one thing in particular that made Rendova such an attractive target to the Allies. On the northern end of the island, there was a 3,400-foot mountain that made for an excellent observation post. They could use it to watch their forces across the channel on New Georgia proper advance on the trail toward Munda Airfield and provide artillery support and covering fire. Coast watchers could also use the mountain as the Allies transformed Rendova into a PT boat base.

But the first Allied men that arrived at Rendova Island on June 30 were met with only mud and plenty of it. The first few days that followed did not go as planned for the 43rd Infantry Division and Admiral Turner's Western Landing Force, as torrential rain bogged down the initial landings. Confusion took over, as supplies were stacked in disorderly piles all over the beach. The cargo got soaked as it was unloaded by hand through waist-deep water. The amphibious assault on Rendova in the first days resulted in a backbreaking, messy disaster. As Rear Admiral Theodore S. Wilkinson once stated, "Unloading is the world-wide difficulty of amphibious operations." This was especially true of the U.S. Navy's Construction Battalions (CBs), or "Seabees." In the Second World War, the Seabees' jobs were to build. Once a new island was invaded, the Navy called in the Seabees with their bulldozers and equipment to start building new forward bases. The Seabees built everything—from the airfields to the hospitals. At one point a group of Seabees on Bougainville adopted the motto: "The difficult we do now, the impossible takes a little longer."

The Seabees of the 24th Naval Construction Battalion accompanied the 172nd Infantry in the landing on Rendova on June 30. Leading the Seabees was Commander H. Roy Whitaker. He later wrote about the unloading nightmare on the beach:

> All day long we sweated and swore and worked to bring the heavy stuff ashore and hide it from the Jap bombers. Our mesh, designed to "snowshoe" vehicles over soft mud, failed miserably. Even our biggest tractors bogged down in the muck. The men ceased to look like men; they looked like slimy frogs working in some prehistoric ooze."

But such was life according to Whitaker: "That's one thing about being a Seabee . . . Mud seems to be our element. When we die, we die in the mud."[3]

Rendova was a hot, steamy wilderness—a paradise for Japanese snipers hidden in the trees towering over the men. With visibility at nearly zero, Allied troops faced a jungle where every single tree had the potential to be a one-man enemy fortress. An onslaught of death could come suddenly from all sides and above, with no warning. However, circumstances would change for Stephen and the crew of the 329. As fate would have it, their landing would happen on a clear, sunny day.

As the 329 loaded up the 169th Battalion at Maquitti Bay, Pavuvu, the Japanese further north were staging a counterattack with "Betty" bombers against the American 43rd Division on stormy Rendova along the beaches. By day's end, twenty-three Seabee sailors from Whitaker's 24th Naval Construction Battalion would be killed from the Japanese bombing raid. Those twenty-three Seabees would not be the last ones to be buried there—three more LCI sailors would join them in the cemetery on Rendova by the sunset of the following evening.

By 5 p.m. on July 3, the 329 and the rest of LCI Flotilla Five were done loading troops. They departed the Russell Islands for their destination of Rendova Island under the cover of darkness. As the 329's skipper W. A. Illing wrote in his war diary, "At last we are doing some good."[4] It was a rough ride through the Blanche Channel. While on the eastern side of Rendova Island, the ships were blown off course and came in west of Tetipari. It was dark and difficult to follow the ships ahead of one another. Some ships passed between Sikuleleki and Boromani Islands. Many of the ships had to go around Sikuleleki Island. When dawn came all vessels were in sight and soon came back into proper formation.

As Flotilla Five reached the northern tip of Rendova Island, they squeezed southward, through the narrow Renard Entrance, sailing into Rendova Harbor at around 8 a.m. on July 4. The 329 beached at around 10 a.m. on Rendova's East Beach, where they unloaded the

last remnants of the 169th Field Artillery Battalion and their equipment. The ramps on each side of the 329 lowered and began to
discharge their temporary guests they'd picked up from Pavuvu the
day prior.

However, enemy reconnaissance planes had spotted Flotilla Five's
approach toward Rendova Harbor in the early morning hours. The
Japanese had other plans for the American landings.

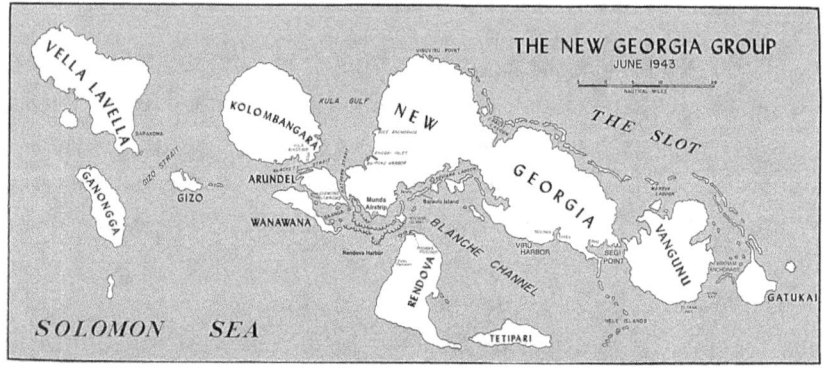

https://www.ibiblio.org/hyperwar/USMC/II/maps/USMC-II-II.jpg

Meanwhile, about four hundred miles to the northwest of Rendova
Harbor, the enemy fortress of Rabaul—the target of Operation Cartwheel—was scrambling in preparation for battle. The mighty enemy
air base on New Britain was the epicenter of Japan's naval and air
power in the South Pacific at that time. Pilots rushed to their planes.
Japanese Vice Admiral Jinichi Kusaka, commander of the *11th Air
Fleet* was determined and desperate to halt the American advance in
the New Georgia Islands. In a last-ditch effort, Kusaka ordered one
last aerial assault in hopes of disrupting American shipping and
supply locations. About one hundred Japanese planes, including
sixteen dual-engine Mitsubishi "Sally" and "Betty" bombers, took off
en route to the southeastern Solomon Islands. Their target: the Allied

landing force beached at Rendova Harbor. An epic encounter was about to ensue.

LCIs 335 (right) and 330 unloading at Rendova on July 4, 1943. (U.S. Navy photo; navsource.org)

By 12:30 p.m. the first group of LCIs, which included the 329, had completed discharging the men of the 169th Field Artillery Battalion and had made their way toward the northern end of Rendova Harbor, where they anchored. Four new LCIs replaced Stephen's group of LCIs on the beaches to disembark troops: LCIs 23, 24, 65, and 63. The LCI 336 joined them as well because it remained stuck on the beach from earlier. The men in Rendova Harbor, including those high and dry aboard the 336, could see a now noticeable aerial battle happening to the north over Munda Airfield, located right across the channel on southern New Georgia proper. American fighters could be seen engaging Japanese planes in dogfights at around 1:30 p.m. At his battle station on the starboard side 20mm gun at the rear of LCI 336, Howard G. Sawyer later wrote of that morning, "As we landed the tide went out and caused a state of confusion." In the distance, Sawyer said, "We saw a fleet of . . . Betty's takeoff from Munda. We thought those Japs were crazy for taking off with all of our air support."[5] The fighting intensified with each passing minute.

The first group of LCIs landing troops and equipment on Rendova Harbor's East Beach on July 4, 1943. (Right to left) USS LCIs 336, 328, 335, 330, 334, and 329. LCIs 327 and 333, not pictured, are believed to be on the starboard (right) side of the LCI 329 (U.S. National Archives photo No. 18-N-52690)

USS LCI (L) 328 (left) and USS LCI (L) 336 (right) landing troops ashore at Rendova in the Solomon Islands, on July 4, 1943. Both LCIs, like Stephen's LCI (L) 329, were assigned to LCI (L) Flotilla Five in the South Pacific. (U.S. National Archives photo No. 80-G-52629)

Over on Baraulu Island, a tiny island located about six miles to the north of where Stephen's ship was currently anchored, Lieutenant Colonel Henry Shafer was in the process of setting up 155mm howitzers with the rest of the 136th Field Artillery Battalion that he commanded. He looked up into the western sky and noticed that around a hundred enemy planes were approaching the umbrella of American fighters that had gathered over Munda Airfield. Like the hunter that waits for his prey to enter a trap, the American fighters had been waiting in the skies above Munda. Waiting to ensnare Vice Admiral Kusaka's air force.

Shafer witnessed the run of the sixteen surviving bombers flying in a stepped-up v-of-vees formation. They were the only ones that managed to make it through the onslaught of dog fighting American fighters waiting for them. The sixteen bombers flew right over Shafer's head, but when they did, he could not see the insignias on the bombers. The bombers curved slightly toward the southeast, then all the way around to the right until they were heading due west toward Rendova Island. Suddenly, Shafer realized the bombers were enemy Japanese. Although he didn't know it at the time, Shafer had a ringside seat to what was about to be an incredible spectacle.[6]

———

As Shafer watched the enemy bombers from Baraulu Island, Stephen and his close buddy, ship's cook Elmo Pucci, were at General Quarters, manning their battle station at the number-two 20mm gun, alertly watching Rendova Harbor. Stephen remembered that Pucci was a "good-looking kid." Since a 20mm gun typically required four men to operate it, each man at their battle station served a different function. Stephen was the gunner, Pucci was the loader, a third man helped aim, and a fourth acted as ammo runner. Lt. (jg) Frank Love Jr. was the officer acting as spotter for their gun. As he stood ready at his gun, Stephen began to hear an eerie sound above him in the distance that he recognized immediately.

"There was this big mountain [on Rendova], but the clouds was really low, so we unloaded the Seabees, and as . . . we started pulling out away—grandpa had real good ears—I could hear a Jap airplane right away because they had a different sound."

And then with suddenness, at long last, after waiting a year and a half since that infamous December morning in 1941 that had jolted the nation to war, the crew of the LCI 329 finally saw the enemy. In perfect formation, sixteen Japanese bombers dipped below the cloud line that hugged the mountain. At about 1:50 p.m., the American artillery on Rendova and nearby Kokorana Island, the LSTs, and the LCIs berthed at East Beach, unleashed a cloud of anti-aircraft fire all at once, aimed at the incoming sixteen Japanese bombers. Every third round in each anti-aircraft gun contained a "tracer bullet"—a shot with a streak of light to assist the gunner with aiming. The sight from thousands of flashing tracers that pierced the sky was quite a sensational sight. The eruption of fire from below blossomed blackly within the bomber V formation. There was no escape from the onslaught of Allied firepower that met the Japanese bombers from all directions.

"When I looked up in the air, here comes sixteen bombers at three thousand feet." Stephen soundly recalled decades later. "I was shooting with the gun at that time, I was on a 20-millimeter, and boy, here they come, sixteen of their big bombers. And when they come over, they was only like about 1,500 to 2,000 feet up, because it was a very low overcast, so they start coming through and right where we dropped all these Seabees off . . . the bombs went down. . . . And that's all grandpa did, . . . I just aim ahead like this, and I just pull, and I just keep pulling." Stephen said, "Out of the sixteen of 'em, we shot fourteen of 'em down." He added, "I must have gotten one or two of them."

Lieutenant (jg) Frank Garfield Love Jr. during World War II. Love, the Executive Officer of the LCI 329 in 1943, spotted for the number-two 20mm gun amidship during the Rendova bombing attack on July 4. (Collection of Mike Love)

And indeed, he did. In the official war diary of the LCI (L) 329, commanding officer Illing stated:

Lieut. (jg) Frank G. Love, Jr., USNR, Executive Officer of this vessel was spotting for Gun No. 2. Tracers of this gun were followed and they were hitting in and near the fuselage of the bomber represented by the numeral Four (4); a few moments after spotting the tracers as described the plane burst into flames and started falling. It is believed that fire from our No. 2 (20MM) brought down this particular Japanese aircraft. Machinist Albert J. Post, USNR, Engineering Officer of this vessel was spotting for Gun No. 3 (20MM) during the above action. Tracers of this gun were spotted hitting in and near the starboard engine of the plane represented by the numeral Two (2). Almost immediately . . . the starboard engine of the . . . plane caught afire and started dropping out of formation. It may be mentioned that this same plane subsequently had its tail blown off by fire of a larger calibre gun than we have aboard, however, it is believed that the fire of our Gun No. 3 started the destruction of this particular aircraft.[7]

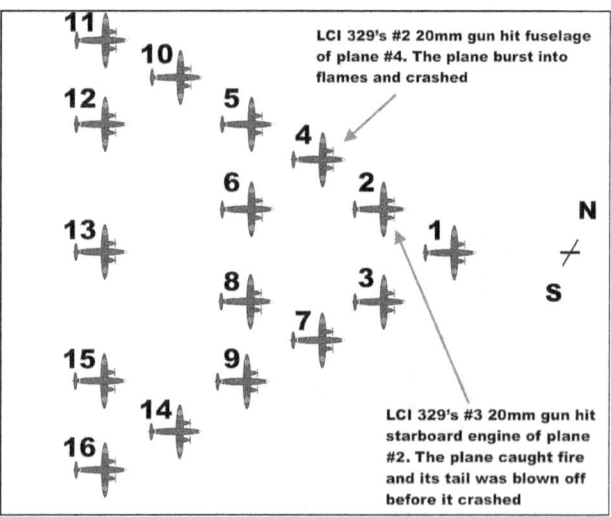

A worms-eye view of the formation of the sixteen Japanese bombers on July 4, 1943. The planes labeled with numerals 2 and 4 were confirmed hit by LCI 329.

"Man, you couldn't miss if you tried. Boy, and that's all I kept going. I'd pull sixty rounds, I told that guy Pucci, I says, 'Keep pumpin' baby, until that barrel gets hot!" Stephen remembered. The thing about 20mm machine guns was that after so many rounds of rapid fire, the gunners were supposed to let the barrels cool off. But LCI gunners didn't shoot according to the book. They just kept blasting away as long as the barrels were white hot.

RENDOVA HARBOR
LCI (L) FLOTILLA 5 -- GROUP 14
4 JULY 1943
14:00 – 14:15

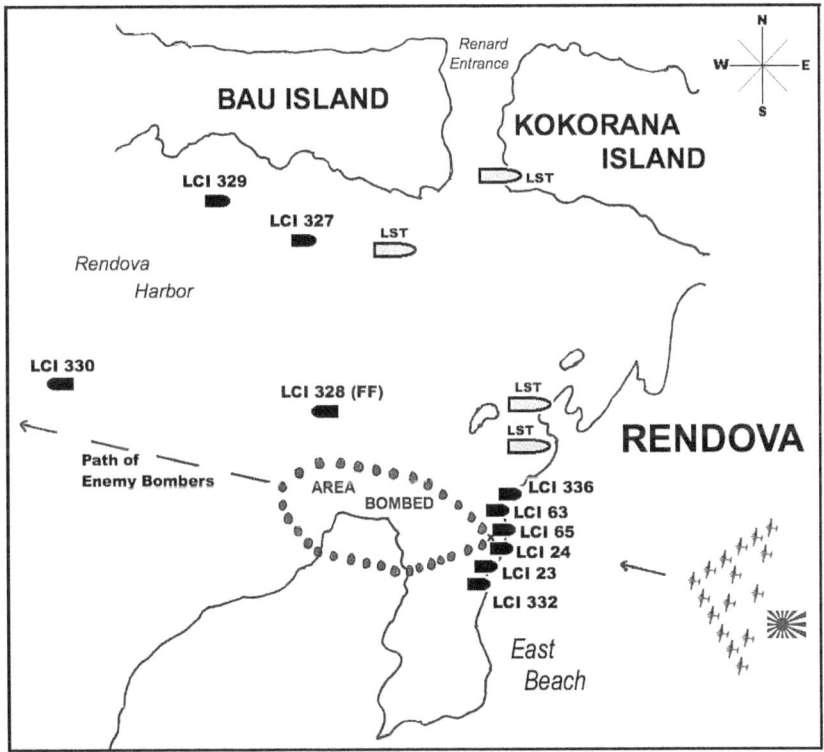

A map of the approximate locations of the LCIs in Group 14 at the time of the bombing on July 4, 1943. LCIs 333, 334, and 335 are not pictured. LCI 335 was exiting Rendova Harbor to the west at the time of the bombing attack. LCIs 333 and 334 had exited Rendova Harbor about an hour before the bombing attack and were anchored near Poko Plantation, about 1.25 miles southwest of Rendova. Scale: 1 inch = 250 yards (approx.)

Lieutenant John R. Powers, a former social worker in civilian life from Cincinnati, Ohio, was aboard the LCI (L) 335, and would live to retell the story of that exact moment with clarity in an article that would run in a year later. "I shall never forget the fireworks that day," he began. "Sixteen Mitsubishi bombers came over and dropped their bombs just after we had retracted. In fact, the bomb pattern fell on the exact spot where we had been beached a few moments before. It was a good old Fourth of July celebration. Planes fell all around us. Our fighters got most of them."[8]

Bomber after bomber fell away in smoldering flames. Tragically, though, due to their low altitude flying, they had managed to release their bombs just before the anti-aircraft fire reached them. The enemy passed directly overhead the LCIs and LSTs anchored and beached in Rendova Harbor. The bombers made a holocaust of the beaches, as the harbor was engulfed in flames. Shrapnel was flying everywhere.

But it was shrapnel from a bomb that exploded between the LCI (L) 24 and LCI (L) 65, that ripped holes in their hulls and would cause the first casualties of LCI sailors in the Pacific War. Three sailors lost their lives, and a dozen more soldiers and sailors were injured from that bomb—which also badly damaged both LCIs, causing them to tilt drastically on their sides.

"The 65, she got straddle-bombed," Stephen remembered before adding, "and that's the ship that I ended up on."

———

Louis V. Plant, a signalman from Detroit standing atop the bridge of the LCI 24, would have that day's events singed into his memory for the rest of his life. Unaware of the role he'd soon assume, he witnessed the deadly shrapnel from Japanese bombs that would take the lives of two of his buddies. The first shipmate was a quiet, older boatswain's mate—Ernest Wilson. Plant witnessed the nightmarish scene and described it years later in his memoirs in vivid detail:

I look forward and I see Wilson lying on his back screaming because the hot deck is burning his flesh. His eyes roll back in his head and he dies from shrapnel wounds. The Quartermaster climbs up on the bridge and says, "Paulson (our radioman) is dead, sir."[9]

Louis V. Plant during World War II. (Collection of Louis Plant)

The quartermaster was referring to Plant's friend, the second man killed, Radioman Mahlon Paulson, who was curious to know what the deafening commotion from all the anti-aircraft gunfire was about. Paulson had left his station in the radio room to investigate. In Plant's memoirs he wrote:

I climb down from the bridge and head for the 'Radio Shack.' Paulson is lying there. He had left his radio to see what the firing was all about instead of hitting the deck and trying to find some kind of cover. A chunk of shrapnel had hit him square in the face and tore most of his head off. The force of the bomb blast had knocked him backward into the 'Radio Shack.' Had he stayed at his radio he might have escaped with just being wounded. . . . I look down on the Port side and see a soldier who has been cut in half above the knees by shrapnel. He says, 'Guess I'll get the Purple Heart for this'...and he dies.[10]

Four other sailors aboard the LCI 24 were wounded: Lt. (jg) Norman Steinert, Coxswain George Dean, Motor Machinist's Mate 1/c James Stewart, and Ship's Cook 3/c Herlof Steffensen. Two soldiers were killed below deck and another was killed while standing on the beach amid the port side of the LCI 24, right next to another wounded soldier who somehow survived the shrapnel. The LCI 23, which was beached next to the LCI 24 on the 24's starboard side, suffered three casualties from fragments of a 100-lb. anti-personnel bomb: Gunner's Mate 3/c Charles Bruce, Gunner's Mate 1/c Paul Warrington, and Motor Machinist's Mate 2/c Emmett Stricklen. Lou Plant could smell the sickening odor of Cordite given off by the bombs that stunk for a long time afterward.

The LCI 65, located on the other side of LCI 24, on the 24's port side, would see one of her sailors perish that afternoon while he was manning the number-four 20mm gun. Fireman 1/c Hurley Christian was struck in the forehead and killed by a flying bomb particle. Aboard the LCI 65 on the number-one 20mm gun, Coxswain W.L. McDaniel and Seaman 1/c Joseph M. Klawon brought down a Japanese bomber according to Major Wilbur E. Brodt, commander of the 169th Field Artillery Battalion, who was observing the action aboard the deck of the 65.

The few remaining Japanese bombers circled around and tried another attack run but were quickly brought down. Twelve of the sixteen Japanese bombers were shot down right there in Rendova Harbor. Other reports later confirmed the American fighters finished off the remaining four bombers as they tried looping around over Munda Airfield to the north. That morning, one hundred sixty-six Japanese fighters had escorted those sixteen Betty bombers from

Rabaul. The attack ended up killing one hundred thirty-seven Americans on the beach that day, and for that, the Allies decimated them in return. By the end of the day, records show not a single Japanese airman survived.[11]

Later that afternoon, signalman Lou Plant and several others were ordered to take Paulson's body ashore and bury him. As Plant recalled in his memoirs, "This was one of the toughest things I have ever had to do, digging a grave for a shipmate with whom you made liberties back in the States. By this time, people are bringing bodies ashore from several ships."[12]

The three LCI sailors were buried together on Rendova Island about a quarter mile east of the landing beaches. Seventy years later, Lou Plant still remembered the suffocating and oppressive stench of death that lingered as the men buried Paulson, Wilson, and Christian. As Ernie Pyle once said, "There is nothing worse in war than the foul odor of death."[13] Plant looked off into the distance and noticed a fire raging from some of Rendova's buildings that had been used as a coconut plantation. A few moments later, a chaplain approached Plant and his shipmates on burial duty and asked them if they were sure that Paulson was dead. The men were positive, they assured the chaplain. The chaplain then asked them if they had removed any personal items from the body, in which Plant replied, "No." The chaplain knelt beside the body and removed Paulson's ring and wallet and placed them in a bag. Plant and his buddies gently placed Paulson's body in a shallow grave and covered it with dirt. They placed palm fronds on top of the dirt, said a prayer, and headed back to the LCI 24.

Because of the damage it sustained from the enemy bombers, Plant remembered seeing his ship listing (tilting) so drastically to port that it had settled on the bottom of the water. Upon returning to the 24, Plant learned that seven or eight soldiers had hurried belowdecks seeking cover when the raid started, as Paulson had. The bomb landed in the water directly opposite the compartment the soldiers were in. The bodies were a mutilated mess. The soldiers' only way to estimate how many of their men the bombs had butchered was by counting helmets and water canteens. The infantrymen carried what was left of their dead ashore in ponchos and shelter halves. As they would later find out, the Japanese bombs had added insult to injury by destroying the turkey allotted to the LCI 24's crew that was meant to be their Fourth of July dinner. Instead, the LCI 24's crew was given canned C rations to eat by the army. Plant remembered that the smell of rotting flesh was so pungent and overpowering that the men could taste it in the food they ate.

But they ate anyway.

The second group of LCIs to land. (Left to right) LCI 23; LCI 24 (listing to port); LCI 65 (listing to starboard); and LCI 63. The LCIs 332 and 336 are not seen in this photo but are beached on the starboard side of the LCI 23 and port side of LCI 63, respectively. Photo taken shortly after the Japanese bombing attack on July 4, 1943. (Courtesy Louis Plant)

————

"They killed a lot of our Seabees," Stephen remembered. They'd died in the mud.

Stephen, Pucci, and the rest of the crew over on the LCI 329 were most fortunate to have narrowly escaped the deadly Japanese bombing attack that afternoon without damage or casualties. While

Lou Plant and his shipmates from the crippled LCI 24 had headed ashore to bury their buddies, the LCI 329 had hauled in their anchor at 6:00 p.m. and exited Rendova Harbor to join a convoy of LCIs returning to Florida Island for further instructions. Their job at Rendova Island was far from over. Enemy bomber and fighter attacks had only just begun. The LCI 329 closed its first day of combat with a bitter taste of the deadly capability of Japanese pilots. Those enemy pilots would only prove deadlier in the days and months to follow.

––––––––

Meanwhile, as Stephen's convoy departed for Florida Island, Lou Plant and the crew of the ailing LCI 24 beached at Rendova were ordered to spend the night ashore until the Seabees of the 24th Construction Battalion could repair their damaged ship. Plant and the crew of the 24 were issued rifles and given instructions on their watch schedule. They waded ashore and climbed into foxholes where they would spend the night on an alien island in harm's way. The news given to the men were that they should be ready for a Japanese invasion force that was expected to come back to recapture Rendova. This was highly disconcerting since there was nothing available to stop an enemy invasion except for some PT boats. Plant remembered how scared everyone felt upon hearing such news.

Suddenly, the LCI 24's crew dug in on the island started seeing naval gunfire off in the distance. Plant, sharing a foxhole with his shipmate Dick Leisenring, clung nervously to his rifle while they both watched the fireworks in the distance. The men were filled with consternation as to what was going to happen next. Plant recognized that the fire they were watching could not have been coming from PT boats. He was sure it was naval gunfire coming from cruisers and destroyers. Little did Plant or his buddies know that they were witnessing the outbreak of the Battle of Kula Gulf. Fortunately for everyone, no harm came to Plant and the men on Rendova that night, but they were all frightened, nonetheless. LCI 336's Howard Sawyer had this to say about that Fourth of July: "Anyone who says they weren't scared to death that day has to [be] lying through his teeth."[14]

After seventy years, Lou Plant still experienced anxiety over what could have happened to the men that day. After the war, he and his shipmate Dick Leisenring stayed in touch and have commented to each other at times over the years how fortunate they were to have survived that night and live to have wives, children, and grandchildren. So many men in their teens and twenties died on that God-forsaken island, only to be buried in shallow, swampy graves far away from their homes and loved ones. They were men whose lives ended before they had barely even gotten started. Plant somberly noted that after that day, the men never again asked when they were going to see some action.

LCI Flotilla Five's first encounter with the enemy had come at last —and on July Fourth of all days. There was no shortage of examples of the men's bravery and fighting spirit in the official reports detailing the events of that afternoon. Lieutenant James McCarthy of the LCI (L) 63 wrote of his men's determination in keeping up their fire even when enemy bombs were hitting close by. Ben Thirkield of the LCI (L) 23 wrote that his crew, most of whom were under enemy fire for the first time, performed excellently. But perhaps the most stirring account of the men's resolve were the words written by Lou Plant's skipper, R. E. Ward—the commanding officer of the LCI (L) 24 who lost two of his own men that Independence Day. In his official action report detailing the events of July 4, 1943, Ward concluded by saying:

> The officers and enlisted personnel fought the ships guns and fires without regard to their personal safety. For a new crew in their first action, they worked quietly, efficiently, and with valor. Individual initiative, courage, and cooperation represented that of the highest traditions of the Navy.[15]

———

"When we hit Rendova, they pulled us in there and we had to stay there for thirty days, and every day, we went through dive-bombing attacks and strafing. And I watched dogfights every day, plain as day." That's how Stephen remembered the month that followed.

"We invaded [New Georgia] and we had to stay up there to make sure that the Army could secure it, if they couldn't secure it, we'd have had to go back in there and pull the guys off. Then what happened was every day . . . this happened for about a month straight—every day you'd look up in the air, you'd see these [dogfighting planes], you know, ours was fighting the Japs. Well, the Japs were pretty good fighters at that time, they had their best naval pilots. And when you see one of theirs go down, one of ours just about went down with him, because that's how good they were. But anyway, we watched that for about, man, almost a solid month."

When Stephen's LCI returned to Florida Island with the rest of their convoy the next morning, July 5, they anchored in Hutchinson Creek at 2:30 p.m. The following morning, July 6, the 329 was again ordered to load up with troops and unload them on Rendova. They headed south with a convoy of LCIs across the Iron Bottom Sound and beached at West Kukum Beach, Guadalcanal, at 8:20 a.m. Captain Heinmiller's 145th Regiment, part of the 37th Infantry Division trudged up the ramps aboard the 329 with their equipment and supplies. After going ashore for further orders, the commanding officers returned to their vessels. The two sections that composed the convoy departed Guadalcanal and headed northwest for Rendova Island. Having been in Rendova Harbor before, Stephen's LCI 329 was chosen to lead the convoy back into the same harbor where the deadly Japanese bombing attack had occurred only four days before. They dropped their ramps and unloaded the troops of the 145th Regiment at 8:00 a.m., before departing again for a pit stop at Florida Island.

By July 9, Stephen and the LCI 329 were back at Florida Island taking on water and preparing to take more troops to Rendova. The Americans' efforts were slowly making progress in the New Georgia Islands. That same day, large naval artillery barrages on Laiana Point (southern New Georgia proper) were helping to clear the way for the advancing 172nd and 169th Regiments' march on Munda Trail.

On July 12, the LCI 329 was back at New Georgia, only this time they were loading troops from the 118th Medical Battalion (part of the 43rd Division) at Oleana Bay, Vangunu Island. Once loaded, the 329 headed for Rendova Island to discharge the 118th Medical Battalion. However, the 329 encountered a potentially fatal miscommunication with the American beachmaster upon their arrival. As the 329 approached Rendova Harbor at 9:21 p.m., they sent a coded message to the beachmaster for permission to enter the harbor and unload their troops:

> *LCI 329:* "*Iceland from Yoke 18. Sausage Watertown. Shall we get veal and unload?*"
> *Beachmaster:* "*Iceland to Yoke 18 – Affirmative.*"

Decoded, the messages simply said:

> *LCI 329:* "*Beachmaster (at Rendova) from LCI 329 outside the Renard Entrance. Shall we get inside and unload?*"
> *Beachmaster:* "*Beachmaster to LCI 329 – Affirmative.*"[16]

The beachmaster replied in the affirmative, giving the LCI 329 clearance to enter. However, for reasons unknown, as the 329 entered Rendova Harbor via the Renard Entrance, the Americans on the coast began to unload friendly fire on Stephen's ship. .45 caliber automatic machine gun fire, as well as .30 caliber rifle fire whizzed past the 329's conning station. After the men ashore ceased firing, the confusion was soon cleared up, and the LCI 329 moored near Rendova's smaller Bau Island at around 10 p.m., obviously without returning fire. They were lucky no one was killed. Another close call.

The next morning, July 13, the LCI 329 beached and discharged the medical troops of the 118th Medical Battalion and their equipment. Upon unloading their troops, officers asked if the LCI 329 could take casualties back to Guadalcanal. The LCI 329 agreed but under the condition that no dangerous neurotics were to be taken. Army C rations were obtained and the casualties came onboard. All manner of fighting equipment was taken from every man and labeled with his name. There were a few rifles and pistols, along with many knives. In all, one hundred seventy-five men came aboard including two war correspondents for return to Guadalcanal.[17] As the LCI 329 was transferring the casualties back to Guadalcanal, the Americans were continuing to make progress on New Georgia proper in their march toward Munda Airfield. The 172nd Infantry Regiment had captured Laiana Point. The next day, July 14, the LCI 329 unloaded their casualties and passengers at Koli Point, Guadalcanal, after giving them back their rifles and knives.

Two days later, on July 16, the LCI 329 was back at Banika, Russell Islands, for more troops. They loaded men from the 148th Infantry Regiment (37th Division) and their supplies for transfer to Rendova, where they unloaded them the next day. Once the 329 completed discharging the 148th Regiment, they returned to Florida Island where they spent the next week and a half resupplying.

———

As Stephen's LCI was loading men from the 148th Infantry Regiment, back in Pennsylvania, at only seventeen years old, Royal Wetzel enlisted in the Navy on July 15, 1943. As previously mentioned, the handsome, Popeye-looking young man would eventually go on to serve aboard the LCI 70 as a cook. Born in Trevorton, Pennsylvania, Wetzel was the "baby of the family," the youngest of five children. His next nearest sibling—his older brother—was eleven years older than him. According to Wetzel, times were tough at home in the wake of the Great Depression. "I got a job when I was sixteen. I went down to Washington, D.C., and got a job driving a bakery truck down there. You were supposed to be eighteen, but I lied about my age." But he added, "That didn't last long. I quit there and then I worked in the mines for a while."

But eventually Wetzel would find himself in the United States Navy where he, like Stephen, also attended bootcamp at Great Lakes Training Station.

On his journey, Wetzel would also follow a similar route that Stephen's LCI 329 took through the South Pacific. Departing from San Diego, "We went over to New Caledonia," Wetzel said. "Then I really don't know how long we were there, but it wasn't very long. Then we were assigned to the LCI 70 . . . we had a very good crew," he remembered. "One guy had a clarinet, and another guy had a fiddle, and we would be singing and carrying on until we got in the war zone." Decades after the war, when Wetzel began attending LCI veteran reunions, he'd earn the nickname "The King of Kazoos" due to his propensity to grab the nearest kazoo and jam out to the first tune that entered his head.

Since the LCI 329 and LCI 65 were in the same LCI Flotilla as LCI 70 throughout most of the war in the Pacific, it meant the LCI 70 would be constantly nearby Stephen's LCIs in the heat of combat to come. Though Wetzel and Stephen would never meet each other, Wetzel would share many of the same experiences as Stephen once he arrived on the front lines. But one similar experience the two did share was—after one particularly long night of drinking—Wetzel, along with his buddies, went out to get a tattoo. On his left arm, Wetzel decided to get a tattoo of a heart with the word "mother" lovingly inscribed on it. He laughed as he recalled his reason for doing it. "I was playing it safe."[18]

———

On the night of July 28, Stephen had yet another close encounter with death off the Russell Islands. At around 10:30 p.m., executive officer Lt. (jg) Frank Love Jr. and Stephen were on duty together in the 329's conning tower. Stephen recalled, "We passed this island, the Russell Islands, and it was pitch black that night, and I remember looking down as I kept steering, and it looked like there was like a fish swimming towards us, and I thought, man that's funny look how straight he is. And I told [Love] I says, 'man look at that, that porpoise is coming straight at us!'"

Love spotted it too.

"It went *fff-eewww*! Right underneath us!" Laughing, Stephen recalled Love turning to him and exclaiming, "'That was no porpoise, that was a torpedo!'" The wake of a torpedo could be seen off the port beam. It was later learned that the torpedo had come from a Japanese submarine in the area. The LCI 329 had missed being hit by a torpedo by only a few feet, and it had its flat bottom-design to thank for that.

"We only drew about a two and a half-foot [draft] forward (front of ship) and like five and a half, six feet aft (back of ship), and for the torpedo to hit you, you got to be down in the water at least fifteen feet . . . so that's why we were lucky . . . we were like a cork."

The month of July was coming to a close. As Stephen and his shipmates remained beached aboard the 329 in the harbor of north-west Rendova, the Americans were in the final days of their grueling, month-long push toward Munda Airfield on New Georgia proper—right across the channel from Rendova Harbor. When nearby, Stephen and the crew could witness the shelling and bombing of Munda. The spectacle of Munda Point's brutal American punishment soon earned a nickname. Men began to refer to the enemy area that endured so much artillery fire and bombings as the "arena."

Munda Airfield—now the "arena," was indeed in ruins, but it was nevertheless still in enemy hands. While Stephen and his buddies were watching the spectacle known as the "arena" in July's closing days aboard the 329, two American infantrymen were awarded Congressional Medals of Honor for their actions during the push through Munda Trail. The first man was First Lieutenant Robert Scott of the 172nd Infantry; he earned his prestigious decoration for his actions near Shimuzu Hill. The second man, Private First-Class Frank Petraca, the battalion medic, was awarded his Medal of Honor posthumously. Petraca had been killed in action by enemy machine gun fire while attempting to clear casualties near Horseshoe Hill. Those two Medal of Honor recipients were among the whole of sleep-deprived soldiers pushing toward Munda enduring a non-stop psychological war with the jungle, the Japanese, and at times, even themselves. Especially at night, panic and war nerves lurked within Munda Trail's rainforest. After one particularly disturbing night where fear had taken over, it sounded like a full blown battle was taking place. The next morning, searchers counted a number of mutilated American corpses, however, not a single Japanese was found.[19]

The Allied forces had been tirelessly fighting to secure the arena. And for that, the month-long push to capture Munda, just like July, was only days away from coming to a close.

4

AUGUST–SEPTEMBER 1943: THE CHAOTIC SOUNDS OF THE SOLOMON ISLANDS

"Enemy planes strike without warning, at any moment, day or night, no matter what you may be doing. Crews who do not watch the sky will not live to see their tomorrow."

– A message from LCI Flotilla Thirteen staff, 1943[1]

"They used to bomb us every day," Stephen began.

The first day of August 1943 fell on a Saturday. It was on this day that Stephen Ganzberger was promoted to Seaman 1/c as the LCI 329 beach-moored near Poko Plantation—a part of northwestern Rendova—as his ship was hiding under the cover of coconut trees near a waterhole. It wasn't exactly the most ideal conditions for enjoying a new promotion. "We was tied up right along by the beach, and the big coconut trees we used to pull 'em over the top of us to try and hide us," Stephen recalled. He and the crew spent the next week hidden in that area while the Americans on nearby New Georgia Island officially secured Munda Airfield on August 5.

Stephen continued: "But those [Japanese], they found us and like every day they'd come down and, *BOOM*, they'd try dive-bombing us. That was kind of scary for a while." The men aboard the 329 were now used to the sound of the General Quarters alarm blaring around the clock. In his book *Brave Men,* Ernie Pyle described witnessing what General Quarters was like onboard an American ship:

> Then all of a sudden . . . General Quarters sounded. . . . The whole ship came to life with a scurry and rattling, sailors dashing to stations before you'd have thought they could get their shoes on. When General Quarters was sounded our sailors didn't get to their stations in the manner of school kids going in when the bell rings. They got there by charging over things and knocking things down. I saw them arrive at gun stations wearing nothing but their drawers. I saw officers upset their dinner and be out of the wardroom by the time the second "beep!" of the alarm signal sounded.[2]

On the morning of August 7, the LCI 329 got underway from their hidden position to change to a new location. They stayed in the vicinity of the waterhole, but they beached just slightly south of their previous position while at General Quarters. Several hours later, a battle between planes could be heard overhead, but due to the clouds, they could not be seen. Suddenly, four Japanese Aichi 99 "Val" dive-bombers appeared out of the clouds. As three enemy bombers flew north toward the PT base at Rendova Harbor to the north, the fourth dive bomber decided to go after the LCIs. The Val changed course slightly, flew overhead, and bombed the waterhole where Stephen's ship had been just hours before. The LCI (L) 328 and LCI (L) 66 were slightly damaged from the bombs. Luckily, there were no casualties aboard the ships.

On August 11, the LCI 329 encountered another case of war neurosis among its men.[3] This time, it was Radioman 3/c Stanley DeJewski, who was transferred to a mobile hospital on Guadalcanal. According to the 329's skipper Illing, "This is our second transfer of cases of war neurosis. Though we need replacements badly and none is in view it is not only better for the man, but also the crew to have him transferred."

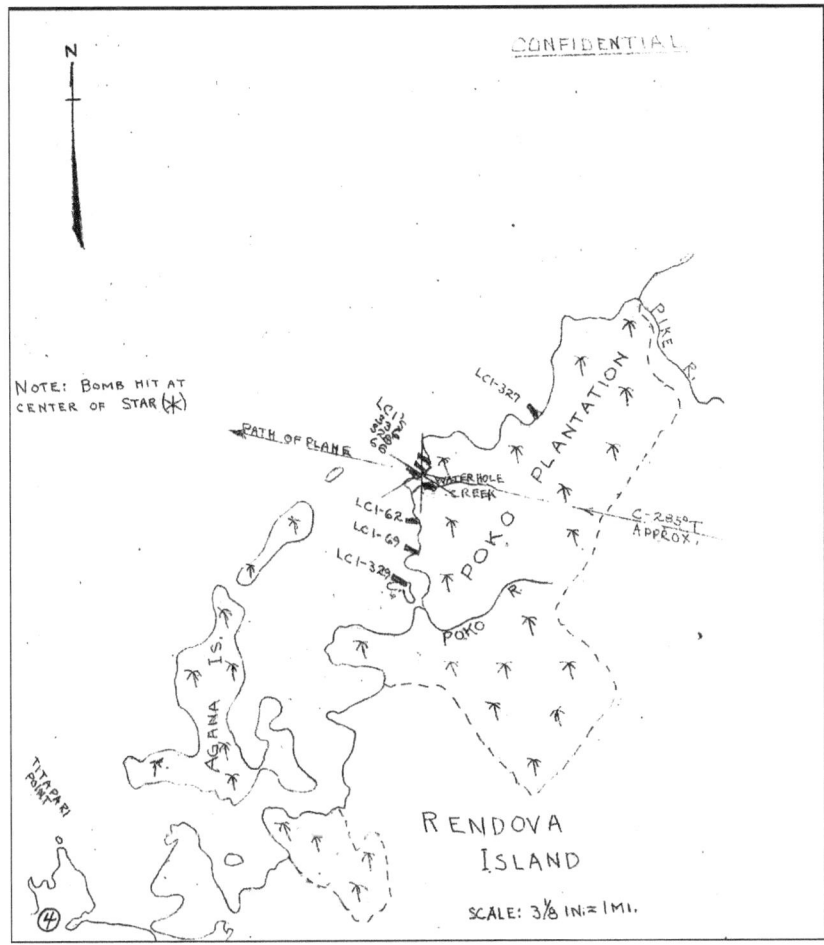

A map of LCI 329's position during the bombing attack near Poko Plantation, Rendova on August 7, 1943. (LCI 329 War Diary, Report of Anti-Aircraft Action by Surface Ships, August 1943, p. 8)

The air attacks from the Japanese were relentless. On August 14, three days later, they struck Stephen's LCI group again at Poko Plantation. At around 11:40 a.m., the LCI 329 fired sixty rounds at three Vals diving at them near the mouth of the river while it was beached with several other LCIs. One of the Vals dropped its bombs in the water, but the bombers managed to escape. Luckily, no hits were registered on any American ships.

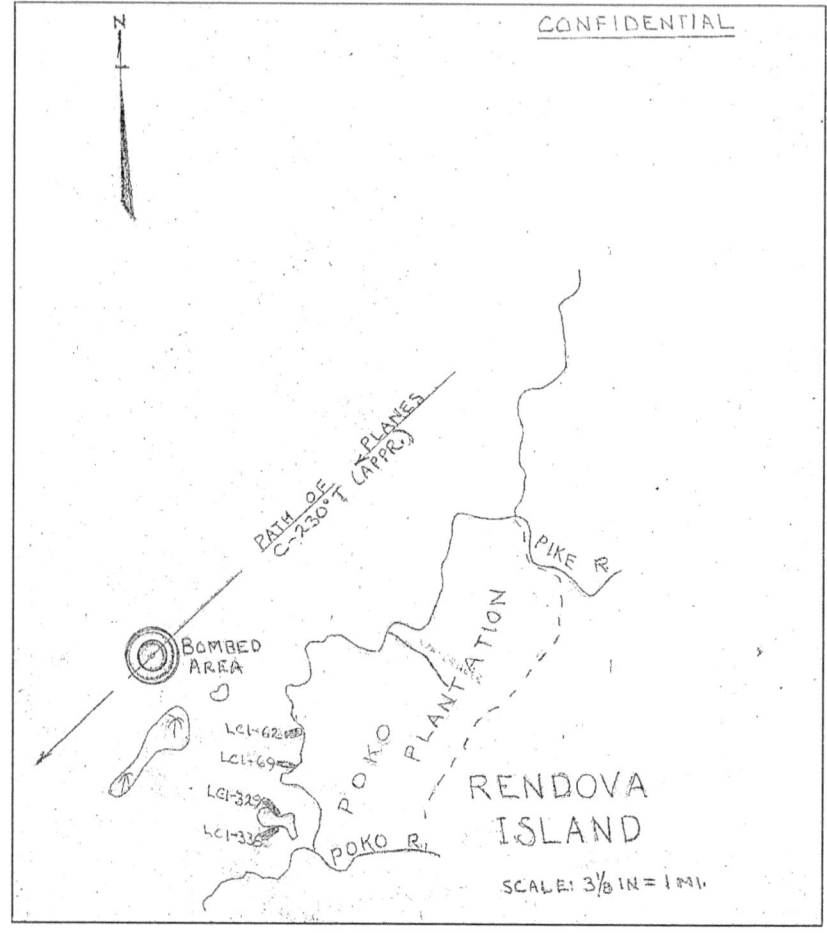

A map of LCI 329's position during the bombing attack near Poko Plantation, Rendova on August 14, 1943. (LCI 329 War Diary, Report of Anti-Aircraft Action by Surface Ships, August 1943, p. 11)

The crew of the LCI 329 also witnessed quite a spectacular event at Munda several days later. While beached at Munda Point in the afternoon, they witnessed about a dozen American dive bombers begin their attack on nearby Baanga Island, located about one and a half miles away from them and about two miles from Munda Point.

Each plane dropped one large bomb with each dive. American artillery soon joined in. Projectiles whistled over the heads of the men aboard the 329 as they smashed into Baanga Island. Though Baanga was bursting with American fire, no Japanese opposition was witnessed. The crew was permitted to go ashore Munda to witness the damage inflicted over the last month by the Americans during their campaign to take the airfield. After seeing firsthand the destruction and punishment done to Munda for about an hour, they returned to the ship and headed back to Poko Plantation. The 329 then headed to Carter City, Florida Island, during the closing days of August. After a hellish thirty days, Stephen and each of his shipmates were also expecting to receive their two-cans-of-beer allotment once they returned to Florida Island but were met with disappointment.

"We was up in Rendova for thirty days. We was tied up alongside of a coconut tree, we were pulling the coconuts over us so the Japs wouldn't see us, but every day we was getting dive-bombed and strafed—that's for thirty days—and then when we come back [to Carter City], we asked [the Navy personnel] for our beer allotment, and he wouldn't give it to us."

Stephen never forgot that injustice.

He and his ship would spend the entire month of September undergoing some much-needed repairs in the southern Solomon Islands.

5

OCTOBER–NOVEMBER 1943: CONSOLIDATION OF THE NORTHERN SOLOMON ISLANDS

"Almost every weapon the Japs have got can reach us on the beaches. We have to take high casualties on the beaches—maybe 40 percent of the assault troops. We have taken such losses before."

– Lieutenant General Holland M. "Howlin' Mad" Smith, USMC[1]

One of the miseries of serving aboard an LCI as a sailor or officer in WWII was the overwhelming absence of almost every necessity and material possession imaginable. Everything was either strictly rationed or out of the question for vessels as small and "expendable" as the landing craft infantry. At times LCIs unfortunately resorted to begging and pleading for survival supplies and nourishment from any nearby naval vessel. The 329 constantly needed engine parts, ammunition, toilet paper, fuel oil, meat, two-year old eggs, beer, and especially fresh water. LCIs also had no surplus water, so these unlucky ships were put on a strict ration. Albert A. Kniewel of the LCI 228 wrote of the situation years later, "Being a small craft we were not able to carry a large supply of fresh meat, which resulted in a diet of Spam 3 times a day 7 days a week." Kniewel added, "I have not eaten Spam since."[2]

A scarcity of laundry facilities also plagued the small ships, so sailors invented their own ways of making do with the situation. They were known to tie their dirty laundry to lines and throw them overboard while underway so they could use the sea to wash their clothes. Even the commanding officers voiced their frustration over the situation from time to time. On October 3, the LCI 329's skipper Illing expressed his displeasure with his current situation while the ship was on duty in the waters off Florida Island. In the official war diary of the 329 he stated:

> While remaining aboard I did some scrubbing of laundry. This idea of having no washing machine aboard, no stewards mate aboard, no laundry facilities ashore, and not being able to have other ships take care of officers laundry is not exactly appealing. During peacetime it would have been considered un-officer like for one to go out on deck and scrub his own clothes. Those of us who commissioned and fitted out these LCI's have learned a lot about "musts" before leaving the continental United States.[3]

Events that unfolded in the month of October led to changes in the grand scheme of the duties conducted by LCIs. About a month before, the Navy had a single Besler Fog Machine installed on each LCI, which would be used in future landing operations to protect and cover the larger American ships with a smokescreen by making clouds of white bellowing smoke. LCIs would lay a smoke screen for carriers, battleships, cruisers, and other capital Navy ships to disrupt the visibility for enemy Japanese looking to attack those targets. This small, yet vital new duty for LCIs would play a critical role in future invasions, especially in the Philippines campaign that would occur exactly one year later in October 1944 when the Allies encountered one of the deadliest and most effective weapons imposed by the Japanese: the kamikaze.

The first few weeks of October were spent clearing the rest of the

New Georgia Islands of remaining Japanese. The next two islands scheduled for invasion were Kolombangara and Vella Lavella. On October 10, the LCI 329 loaded troops of the 1st Battalion, Fijian Infantry Regiment at Hutchinson Point, Florida Island and got underway. At 11:45 a.m. the following morning the 329 and several other ships proceeded in a single column through the passage between Arundel and Baanga Islands and on through the Diamond Narrows, then to Vila, Kolombangara. All ships beached and the 329 disembarked eight officers and one hundred eighty men of the 1st Fijian Battalion. Decades later, Stephen described those men from Fiji saying, "The guys there on [Fiji] average six foot. Every one of them that we picked up . . . they were what they called 'knife fighters.' They'd go out and the only thing they carried with 'em when they went was a knife. And they was just slittin' throats." He added, "and they were cool."

Afterward, per the usual routine, the 329 returned to their home base at Hutchinson Creek, Florida Island, where they spent several days refilling supplies, taking on fresh water, and making welding repairs.

On October 15, Stephen's ship headed for Kukum, Guadalcanal under orders from Task Group 31.1 on a secret dispatch. There, they embarked one warrant officer and ninety-one marines. After joining up with a task unit comprised of the LST 399, LST 70, and APc 50, the LCI 329 headed once again for Munda at a speed of about nine knots. They arrived the next morning, on October 16, where they disembarked more Army troops, and later that afternoon, picked up more infantrymen from the 25th Division on Olson's Beach. The 329 delivered them back to Kukum, Guadalcanal, the next morning. Stephen and his ship would spend the rest of October in the Govana Inlet at Florida Island grinding away at their usual routine.

Though the closing of October was rather uneventful for Stephen and the 329, it was quite the opposite for the rest of the amphibious ships gearing up in Hutchinson Creek nearby. Starting on October 27, the Allies launched the beginning of their next big seizure in their island-hopping strategy in the Pacific. Dubbed "Operation Cherry

Blossom," the invasion of the main island of Bougainville was scheduled for November 1, but before the Allies could invade the mainland, they first needed to land the 3rd New Zealand Division on two small islands just to the south, called the Treasury Islands.

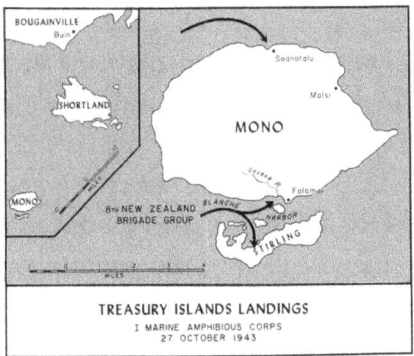

Bougainville's Treasury Islands are made up of Mono and Stirling Islands. Map from *Isolation of Rabaul: History of U.S. Marine Corps Operations in World War II, Volume II* (1963), by Henry I. Shaw, Jr. and Douglas T. Kane, p. 190.

———

A little over three hundred forty miles to the northwest of where the LCI 329 spent the closing days of October, the U.S. Navy was about to demonstrate a newly designed weapon and Stephen's LCI Flotilla Five was to be directly involved. A brilliant and innovative idea had occurred to the Navy leadership to convert LCI (L)s into gunboats to be used during the initial first waves during landings. Since their flat bottoms allowed them to get close to the beaches, if the LCIs were more heavily armed, the Navy gathered they could be used as gunboats against the Japanese, who would be dug in under heavily fortified protection like concrete pillboxes. These heavier armed LCIs could even maneuver in between tiny islands and through rivers, and they were about to be put to the test.

From early to mid-October 1943, four LCI(L)s—21, 22, 23, and 70 from LCI Flotilla Five—were converted into gunboats at Noumea, New Caledonia. Those four ships, hereafter referred to as LCI (G)s, or

gunboats, barely made it back in time for the operation, arriving at
Hutchinson Creek, Florida Island, on October 23, close by Stephen's
LCI. The LCI (G)s had upgraded their armament and weaponry. The
LCIs would no longer transport troops; their new mission would be
to provide close-in fire support as the troop-carriers landed soldiers
and Marines on enemy-held beaches.[4] According to official Navy
reports, the four gunboats each added a large three-inch 50 caliber
heavy machine gun, which was mounted on a platform above and
between the forward bulwarks. The first two LCIs converted to
gunboats were the LCIs 22 and 23. Charles Ports, who served on LCI
23, which was one of those first two gunboats, remembered the
conversion and being assigned to the gun crew of the new heavy
machine gun:

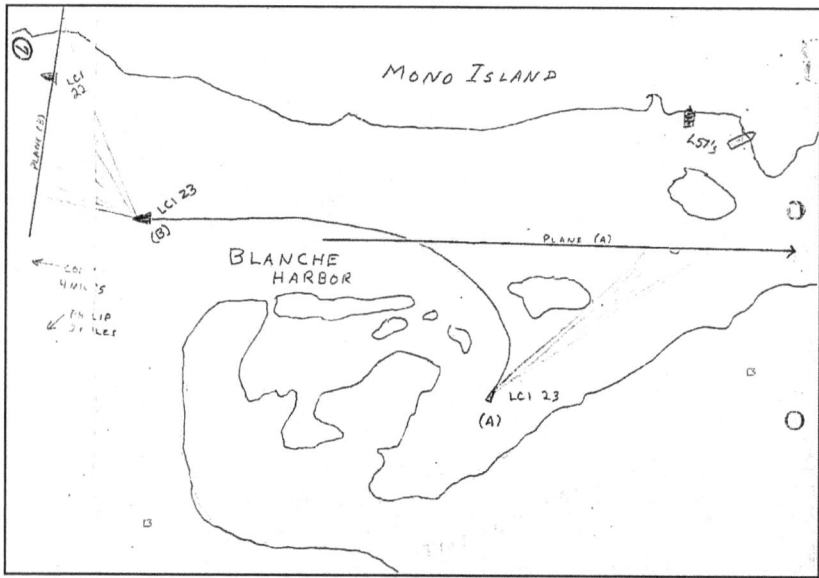

A map sketch from the action report of USS LCI (L) 23, dated October 27, 1943, at
Treasury Islands (p. 33). LCI 23 was one of the first two LCI gunboats to officially
see combat in an amphibious assault.

On 27 October 1943, I witnessed my first combat at Blanche Harbor,
Treasury Islands, as the first loader of the 3" 50 gun, the combat
station I held until I departed the vessel. We were close enough to

the beach that small arms fire was ricocheting off our structure and on our strafing; one could see bodies falling from the trees. Two LSTs were beached and were unloading under mortar fire when we moved into a visual position and placed several rounds of 3"50 into an area from where smoke was coming. We hung around a bit but on leaving the area, the mortar attack resumed so we moved back and placed several more rounds into the suspected position after which no further action was detected.[5]

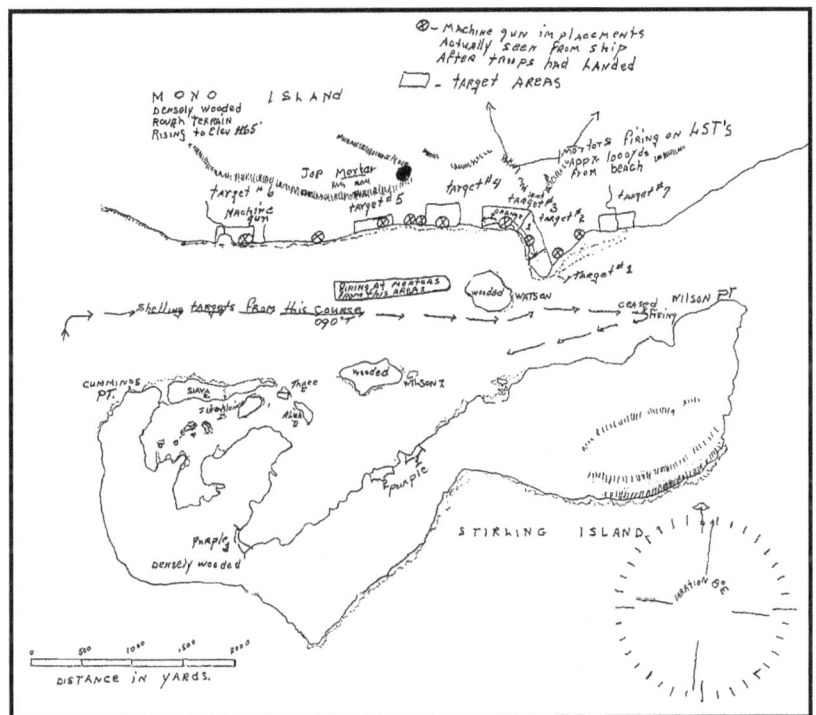

A map sketch from the action report of USS LCI (L) 22, dated November 5, 1943 (Enclosure A). Note the "X" targets show the Japanese machine gun emplacements on Mono Island.

The next morning, Charlie Ports's commanding officer, Ben Thirkield, typed up his action report of the LCI 23's mission at Mono and Stirling and sent it off to the commander of his Task Group 31.1, G. H. Fort. Thirkield's report—filled with details of the LCI 23's heroics of rescuing a disabled landing craft, vehicle, personnel (LCVP), engaging two enemy planes, and covering the distressed LSTs on the beach by blasting enemy gun emplacements—also detailed the success of the gunboat despite the crew's inexperience using the new heavy gun. The report eventually made its way to the commander of the U.S. Third Amphibious Force, Rear Admiral Theodore S. Wilkinson. Wilkinson read Thirkield's comments near the conclusion of the report that stated, "As the first operation for the converted LCI(L) type gunboat, this section appears . . . to have demonstrated this great potential value of heavily armed small craft in supporting landing operations against hostile beaches." He went on. "Some recommendations regarding their use have suggested themselves in the meager light of this one experience." Thirkield then suggested what became perhaps his most important contribution to the war effort. He recommended:

(a) That support gunboats accompany assault waves all the way to the beach, turning away only in time to avoid beaching themselves. This would provide flank cover for the assault wave to the last possible moment, instead of exposing the flank as occurred in this action . . .

(b) That support gunboats whenever possible, be free of troops, in order that they may retain complete mobility throughout the operation.

(c) That liaison be established between gunboats and shore fire control parties to enable gunboats to provide intelligent supporting fire after landings.

(d) At the first opportunity, and at every opportunity thereafter, the gunboats should be given target practice against shore, surface, and air targets. This is considered essential if these craft are to attain their maximum potential effectiveness. It is also strongly urged that an officer-instructor be assigned to these vessels temporarily to conduct training of officers and men in the use of the 3" 50 caliber recently installed. Such training and assistance is considered most

urgent by this writer, because of the complete lack of experienced officer or enlisted personnel.[6]

Wilkinson forwarded the recommendations to Admiral Halsey's
Chief of Staff, Robert Carney, but before he did, Wilkinson added his
own endorsement, commenting, "The LCI(L) 23 along with LCI(L) 22
were the first two LCI gunboats to be employed in combat in this
area. Both ships proved highly effective as support vessels for this
particular operation, and, it is believed, can be gainfully employed in
future amphibious operations." Wilkinson added, "As a result . . . four
additional LCIs are now being converted to gunboats and will be
employed not only as support boats for landing operations but also
for supporting PT boats in coordinated anti-barge missions."[7] By
December 5, Carney himself endorsed Thirkield's report after
reading it. Impressed, Carney agreed and declared that Thirkield's
recommendations "appear to be sound in the light of experience to
date."[8] A massive conversion effort was officially underway.

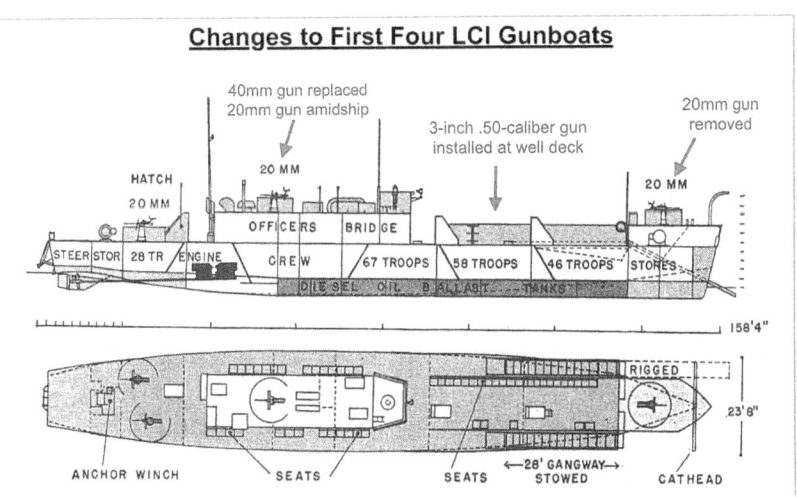

A blueprint of the original LCI concept with additional labels. In future
gunboats, several of the 20mm guns were replaced with heavier three-inch 50
caliber and 40mm guns, and six .50 caliber machine guns were equipped. Photo
from Daniel Barbey's *MacArthur's Amphibious Navy*, p. 361.

On the first day of November 1943, the men of the 3rd Marine and 37th Infantry Divisions entered Empress Augusta Bay, and invaded Bougainville Island. The capture of the last, largest, and most crucial island of the northern Solomon Islands was under way. Meanwhile to the south, beached in Berth 14 of Hutchinson Point, Florida Island, it was just a routine day for Stephen and the crew of the LCI 329. But they wouldn't be out of the action for long. The next day, the crew of the 329 would install two additional .50 caliber air-cooled Browning machine guns for their coming action.

Shortly after noon on November 9, the day after their one-year anniversary of being commissioned, the 329 beached at Kukum Beach, Guadalcanal, and embarked one hundred sixty-three men and seven officers, including men from New Zealand's 29th LAA Regiment. They would be part of the 4th Treasury Echelon, which consisted of six LCIs, three LSTs, two destroyers, two minesweepers, and a coastal transport vessel. By the evening, the ships were ready to depart Guadalcanal. Under the command of Commander W. L. Dyer aboard the destroyer USS *Sigourney,* the 4th Treasury Echelon got underway and sailed in formation to southern Bougainville. Stephen and the crew of the 329 were now heading to their next invasion: Stirling Island, Treasury Islands.

At around 1:10 a.m., the ships of the 4th Echelon sounded General Quarters. A Japanese float plane suddenly appeared from out of the early morning sky. It dropped two bombs between the LST 166 and the coastal transport vessel but did not hit or damage either of them. The Japanese plane veered off so quickly that none of the ships had a chance to even fire a single round. Luckily, that was the first and last threat to the convoy.

"There was a lot of excitement through here," Stephen remembered when recalling Bougainville. Although the invasion of Bougainville began in November 1943, the Allies would continue their fight for island for the remainder of the war. Starting on November 1, the Allies would be locked in a struggle for the island over the next two years, until Japan's surrender in August 1945. "I forgot how many times I landed on Bougainville with troops," Stephen reflected.

At 7:05 a.m., on the morning of November 11, the LCI 329 plodded through the stormy morning tide of Blanche Harbor and beached at Purple Beach 2, Stirling Island, just to the south of the main island of Bougainville. Stephen and the crew of the 329 smashed onto the sands of Stirling Island and dropped the ramps.

"Our ship went right to the beach, and on each side of us were two big ramps. We dropped 'em off and the soldiers would go off," Stephen declared. They unloaded all the troops and supplies in just under an hour. The 329 joined up with a larger convoy, where they cleared Blanche Harbor and headed back to Florida Island. The Treasury Islands were officially secured the next morning, November 12.

Just over a week later on November 20, as Stephen and the 329 were spending their morning at Tulagi, another event was happening nearly eleven hundred miles to the northeast that would change the course of America's amphibious warfare in World War II.

———

Up until this point in the war, General Douglas MacArthur was commanding the Southwest Pacific Area in New Guinea with the Seventh Fleet, while Admiral William Halsey was commanding the South Pacific Area in the Solomon Islands with the Third Fleet. As discussed, it was a two-pronged, island-hopping campaign, fought simultaneously as the two armies drove toward Rabaul, New Britain. But on the morning of November 20, everything changed. Under the overall command of Admiral Chester Nimitz, Vice Admiral Raymond Spruance now commanded the Fifth Fleet aboard the heavy cruiser USS *Indianapolis* in the first offensive of the Central Pacific Area. A third prong in the attack against the Japanese in the Pacific Theater was opened. Their first target was a tiny island named Betio in the Tarawa Atoll.

———

In military terminology, the day a force strikes a new island or country is called D-Day, and the time it hits the beach is H-hour. On D-Day at Tarawa, it was a dangerous coral reef surrounding Betio Island that held up scores of Marines as they attempted to land on the beach at H-Hour. The razor sharp coral caused the American landing boats to grind to a halt, and the Marines were gunned down as they waded ashore in waters chest high. The Japanese defenders had been ready and waiting for them. The Battle of Tarawa would be the first time in the war in the Pacific that the U.S. forces faced massive Japanese opposition to an amphibious landing. All previous amphibious invasions had met little or no initial resistance to the American soldiers and Marines landing on the beaches, but that changed at Tarawa. The battle revealed the dangers of Marines' complete exposure as they approached enemy-infested beaches.

Prior to the battle, as the large Allied ships with the heaviest firepower pounded the island unmercifully from far offshore, the tiny personnel craft loaded up with Marines who were preparing to invade. As the American battleships bombarded the enemy-held island, the Japanese hunkered below underground, usually under the protection of heavily fortified bunkers made with concrete, trying to survive the siege of heavy screaming shells. However, once the small landing craft full of invading Marines headed toward the island and crossed the line of departure, the heavy firepower from the large ships had to stop, because they couldn't risk killing their own men. This critical few minute period, referred to as the "lull period," allowed the Japanese just enough time to reposition, resupply, and rearm themselves. If there was no continuous supporting fire, by the time the small landing craft hit the beach only minutes later, it was havoc for the troops. At Tarawa, there was no protective fire other than the individual gunners armed on the tiny Higgins boats and amtracs[9]—and they were no match for the Japanese dug into Betio Island on D-Day.

———

"A million men cannot take Tarawa in a hundred years."

That was the declaration boasted by Japanese Rear Admiral Keiji Shibasaki, the commanding officer in charge of making Tarawa one of the most fortified islands in the history of warfare. In the months leading up to the battle, Shibasaki transformed the one-square-mile island into an underground fortress of interlocking caves, pillboxes, and bunkers, all protected by inches of concrete, sand, logs, and rock. Even under the heaviest American bombardment, the Japanese were safe. Under his command, the forty-five hundred Japanese defenders were well-supplied and well-prepared, and they were determined to hold Tarawa at all costs.

During the initial invasion on November 20, Shibasaki's men watched as the American landing craft got hung up on the reef inside the atoll and were sitting ducks to the Japanese guns that tore into them. Only the smaller amtracs were able to traverse across the reef, and of those that didn't get destroyed, the Marines were forced off their crafts and into the lagoon. There, they waded ashore in chest-high waters, slowly making their way to the enemy beach still seven hundred yards away—through an onslaught of bullets, barbed wire, and underwater obstacles designed to trap intruders. It was a nightmarish slaughter.

"Because the naval bombardment had ceased," Japanese Warrant Officer Kiyoshi Ota wrote later, "we took advantage of the opportunity to position ourselves carefully, to emplace our machine gunners and riflemen in the best spots, and to rearrange our ammunition supplies to the best advantage."[10]

As a result, the fight to capture Tarawa ended with over three thousand casualties, including over a thousand lives lost for both the Navy and Marine Corps.

———

As the U.S. military does, it observed the battlefield horrors and adapted. As the former commanding officer of LCI 345, Charles P. Grow, remembered it, "The Marines had just taken the island of Tarawa, at great cost. It was apparent that, if the Pacific mopping-up operations were to continue, the amphibious forces would require better naval support."[11] Tarawa had exposed a fatal problem that needed to be solved to prevent more unnecessary bloodshed and men from being killed, and the U.S. Navy decided that their approach to amphibious warfare needed to change immediately. All focus shifted to eliminating the lull period and it was decided that a small, flat-bottomed amphibious landing craft already involved in the fight was the solution to its problem. After all, the success of the four recently converted LCI gunboats in the invasion of the Treasury Islands at Bougainville had already made itself known to Navy high command.[12] Starting with the orders of Wilkinson and Halsey, the Navy's solution was to upgrade the armament of the LCIs and send them into battle with far more firepower, joining the first waves to prevent any lull period from happening. The LCI was the missing piece it needed.

However, there was a consequence of adding more firepower. Those LCIs converted to a heavier armament would need their ramps removed to make room for the larger machine guns and more crewmen that would be required to operate them. LCI gunboats—or LCI (G)s—would therefore no longer have the capability to land men on beaches. Their primary function would be shifted from troop delivery to troop protection. LCI (G)s were to advance alongside the invading troops and provide continuous covering fire for them, as LCIs were among the only ships that could get close enough to the beaches to eliminate this lull period because of their flat bottoms. LCIs were about to become more heavily armed with larger crews for the duration of the war, thanks to the horrendously deadly lessons learned at Bougainville and Tarawa.

———

The way an LCI was armed determined its role and responsibility in an invasion, and its role, therefore, determined the ship's new designation. Meaning the LCI (L) would get a new designation once it was converted. Instead of being known as a Landing Craft Infantry (Large)—or LCI (L)—it would be renamed according to its type of armament. For example, two months after the Battle of Tarawa, when Stephen transferred from the LCI 329 to the LCI (L) 65, the 65 was eventually converted into a gunboat, being renamed LCI (G) 65—replacing the "L" (Large) with a "G" (Guns).

USS LCI (R) 73 underway as a unit of the Seventh Amphibious Group, date and location unknown. The LCI 73 was one of many LCIs that were converted into a rocket-firing LCI. Notice the rocket racks installed on her port side. (navsource.org; U.S. National Archives photo)

Other LCIs had rocket launcher racks added in place of their ramps on each side of the ship. These LCIs were capable of firing hundreds of Mk.7 rockets like bazookas. They would provide fire support for landing troops, and each was known as a rocket ship—or LCI (R). When the Elsies were converted into rocket-firing ships, the two landing ramps were removed from the bow, and metal racks—called launchers—were welded to the steel decks on both sides of the the well deck. The launchers used a simple mechanism, an electrical firing system connected with a control station that was installed on the bridge.

Another unique variation of the LCI was the LCI mortar ship—or LCI (M). These were armed with mortars installed on each side of the ship. During invasions, these LCI (M)s could fire on targets close by on land that posed a threat even though those targets were not directly facing the beaches. The mortar ship LCIs could loft 4.2-inch CML mortars (a type of launched explosive) over a hill where Japanese were often hidden on the other side, not facing the landing beach. The ability for the men of the first waves to have rocket blasts and lobbed mortar fire covering their landing was invaluable and could most certainly be credited for saving potentially thousands of lives.

Historian and author William L. McGee summed it up well when he wrote, "The birth of LCI (Rocket), LCI (Gun), and LCI (Mortar) ships was due to necessity. It was observed that the last four minutes of the approach of the first wave of troops of an island invading force were sitting ducks. This was the most critical time of the whole operation. The Battleships and Cruisers laying out on the horizon and the Destroyers closer in had to stop firing due to the proximity of the first wave to the beach," but he continued, "we needed a close-in support ship that could go in ahead of the first wave to cover the critical four minutes before the first wave hit the beach." McGee went on to say that the subsequent success of the two LCI (R)s in January 1944 at Kwajalein Atoll "was so great that over 100 LCIs were soon converted to rocket, gun, and mortar ships. From then on, many of the LCIs were used only for close-in fire support for our landing forces."[13]

As the men approached the landing beaches in future invasions, the LCI (G)s, (R)s, and (M)s would open up on the Japanese with bullets, rockets, and mortars. The LCI gunboats would shred away at the Japanese with their heavy machine gun fire. The LCI rocket ships would create a devastating wall of fire and explosions from the Mk.7 rockets that could eradicate Japanese hiding in pillboxes and concrete bunkers. And the LCI mortar ships could obliterate hidden threats behind hills that no other ship could hit. In late 1943 and early 1944, it became common for most of the new LCI (L)-351-class round-conn LCIs to be constructed from the very beginning as gunboats, rocket ships, or mortar ships. These three types of LCIs would help eliminate the existence of a lull period in the invasions to come.

USS LCI (R) 72 firing rockets at Morotai Island on September 15, 1944, after having rocket racks installed. (U.S. Army Signal Corps photo No. SC 194499 by C. Perry, U.S. National Archives)

———

Japanese Admiral Shibasaki was wrong. It didn't take a million men a hundred years to take Tarawa. It took the United States Navy and Marine Corps seventy-six hours and an invasion force of around 53,000 men. The dead bodies of young men from all across the United States littered the island, the beaches, and the water. They continued to wash up onto the beach in the days after the invasion. Red Beaches 1, 2, and 3 were literally turned red as a result of all the blood that had been spilled. But their deaths were not in vain, as Tarawa was declared secure on Tuesday, November 23, 1943.

That same evening, as Stephen and the crew of the LCI 329 enjoyed a Thanksgiving turkey dinner (two days early) while beached at Berth 10 at Hutchinson Point, Florida Island, they received some unexpectedly great news. The 329 had just gotten orders to depart for a recreation trip to Sydney, Australia. Thoughts consumed by loved ones back home suddenly shifted to thoughts of the land down under. The crew was elated as they departed the Solomon Islands on November 26. From a coastal transport nearby, Stephen's ship picked up one officer and fourteen enlisted men who made the trip with them as passengers. The 329 would be forced to make a stop in New Caledonia first, but Stephen didn't care. *It's about damn time* he thought. He and the crew would finally be granted liberty after nine months of nonstop action in the hellish Pacific.

6

DECEMBER 1943: DOWN UNDER

"Women and girls lined the way as we went through small stations, all waving and blowing kisses."

– Sid Phillips, *You'll Be Sor-ree! A Guadalcanal Marine Remembers the Pacific War*

After making a brief stop in New Caledonia to pick up six more sailors for temporary passage to Australia, Stephen and the LCI 329 entered Sydney Harbor on December 6 at 8:35 a.m. Though Stephen had no knowledge of it at the time, he and his ship, over the course of the next several days, would be moored right next to the very ship he'd soon be transferring aboard in a little over a month—USS LCI (L) 65. Stephen had been striking (training) for the job of quartermaster, and the 65 would need one in mid-January 1944. This would be Stephen's last liberty with his buddies, as these were his final days with the "plank crew" of the 329.

As the 329 sat moored in Berth 10 of Woolloomooloo Bay along-side the LCI 331 (to port) and LCI 65 (to starboard), Stephen looked around at the strange new continent. It was exactly two years since the Japanese attacked Pearl Harbor, and the nineteen-year-old from Wyandotte found himself a half a world away from his hometown in yet another place war had carried him that he'd never been before. He and the crew had long anticipated this well-deserved liberty. But given the length of time Stephen had been serving on active duty on the front lines and given the length of time the men were allowed liberty (only a week), one couldn't help but agree with him when he said, "It wasn't much."

Decades later when Stephen's wife, Patricia, of sixty years overheard Stephen recall that he only got one week of rest throughout his entire Navy career aboard an LCI for over two years of combat duty, she turned and exclaimed in utter disbelief, "That's it?" Again, he simply replied, "Yup."

On December 9, the 329 left Woolloomooloo Bay and entered drydock at Morts Dry Docks in Walsh Bay, Sydney Harbor. The 329 underwent minor repairs as the crew spent the next two days scraping and painting the hull in drydock. On December 11, the 329 undocked and moored alongside the LCI 65 once again. The 329 would remain closely moored alongside Stephen's next ship for nearly a week. Charles Ports of the LCI 23 brightly remembered, "We had a good time and ate well while visiting Sidney."[1] While the young men roamed the streets, they were greeted by the grateful locals of Sydney, especially young ladies, with smiles, winks, and banners and pennants displaying "Good on you, Yank! You saved Australia!"

Crewmen of the LCI 226 also learned what a friendly people the Australians were. They soon discovered the girls whistled back twice as loud as the Americans whistled at them. In one instance when the LCI 226 first approached Woolloomooloo, skipper Henry T. McKnight recalled noticing his ship's signalman atop the conning tower frantically waving signals in the direction of a park. The signalman explained he had heard rumors that many Australian girls remembered semaphore from their Girl Scout days and he was trying to fix up dates.[2]

While Stephen was in Sydney, he picked up a newly updated admiralty navigation manual, a book that was used to guide Australian officers training in "His Majesty's Fleet." Though the brown cover has worn and stained over the years, the book still remains in Stephen's family today.

The 329's sojourn in Sydney came to an end on December 19, when they departed Woolloomooloo Bay at 2:23 p.m. and joined the LCI 333 on their journey back to New Caledonia, where they would arrive three days before Christmas. On Christmas Eve, the 329 received 7,000 gallons of fresh water and twenty-six marines aboard for passage to Solomon Islands. The crew of the 329 and the twenty-six marines spent Christmas Day moored alongside the LCI 66 in Great Road, Port Noumea, New Caledonia. On the night of December 27, under the cover of darkness, the 329 departed New Caledonia for Guadalcanal, where they arrived three days later on the evening of December 30.

New Year's Eve 1943 brought both joy and sadness. On that day, Stephen was finally promoted from a first-class seaman to a Quartermaster 3/c. For the last six months, Stephen had been striking to be a quartermaster. When recalling the duties of a quartermaster, Stephen reflected, "I handled all the navigation, like correcting the charts, chronometer, yadda-yadda." The quartermaster also handled the steering of the ship. As a quartermaster, Stephen's location of duty while on watch was inside the wheelhouse within the LCI's conning tower at the center of the ship. From then on, when the men manned their battle stations during General Quarters, Stephen would remain in the conning tower to navigate the ship, instead of running to his 20mm machine gun.

But on that same day the crew also received news that their beloved Executive Officer, Frank G. Love Jr., would be leaving the ship to become the permanent commanding officer of the LCI 64. As the young men said their goodbyes to Love, Stephen wondered if he would ever see him again.

———

Meanwhile on the front lines, the 1st Marine Division began their invasion of New Britain Island at Cape Gloucester the day after Christmas 1943. The Old Breed's objective was to establish a beachhead and capture the airfield. Then they'd move eastward, making their way across the entire island, eventually neutralizing the once mighty Japanese airbase of Rabaul. The airfields from which the Betty and Sally bombers took off from that had bombed Stephen's LCI group at Rendova were now reduced to rubble. The Allies were strangling Rabaul and Operation Cartwheel was inching closer to its conclusion. Alongside the Marines during the invasion were the LCI rocket ships making their battlefield debut. Within the smoke and fog, renowned war correspondent John Hersey was present for the invasion aboard the LCI (L) 226 and later described the brand-new rocket ships' devastating effect in an article published a few months later. The 226's engineering officer, who'd witnessed the bombardment from aboard the LCI rocket ships, stated in the article that it looked like the LCIs had sunk the entire beach.[3]

Like the success of the first four LCI gunboats at Treasury Islands two months before, the success of the LCI rocket ships at Cape Gloucester quickly made itself known to Navy leadership. Soon the Navy would order far more LCIs to be converted into rocket LCIs for future invasions. Many of them would be ready just in time for the Marshall Islands campaign the following month. By the end of 1943, the Amphibious Force had churned out tens of thousands of brand-new LCIs, LSTs, LCMs, LCTs, LCP(R)s, LCVPs, and LSDs.[4]

When the war began most of these amphibious ships had not even been designed. But now, every one of them was becoming an integrated Navy vessel, each with its own angry little personality. Collectively, up until this point in the war, they had already made a lasting impact all over the world. And soon they would make victory possible in the decisive battles to come.

7

JANUARY 1944: A NEW SHIP

"It almost became like a sport. It's an awful thing to say, but that's how it was. Your enemy was your enemy. Like it would be for a football team, consequently, you all went in united. Your crew pulled as one."

– Eddie Benoit, LCI 74

Stephen, now a Quartermaster 3/c, and the crew of the 329 began New Year's Day 1944 beached in Berth 10, Hutchinson Point, Florida Island. The crew moved locations shortly after 8:30 in the morning, then spent the rest of the day overhauling, replacing, and repairing their main engines and installing other equipment at Carter City. They would spend the next five days moving around Carter City testing out their repairs and installations. However, on January 6, a most unfortunate event set the 329 back even further behind schedule. Shortly before 3 p.m., the LCI (L) 443 accidentally rammed the back of the 329 as it was attempting to come alongside Stephen's ship. The collision damaged the 329's chocks, stanchion, and ready service box on the starboard side. The skipper, Illing, dove into the water to check the ship's two screws (propellers), where he found a two-inch line of rope wrapped around the shaft. He cleared the screws of the line as the ship got underway. Two days later, the 329's crew finally completed overhauling, replacing, and repairing their main engines. They also finished their installation of other equipment.

Then Stephen's big day finally arrived. On January 16, 1944, he gathered up his belongings in his sea bag and said goodbye to Elmo Pucci and the rest of his buddies in the 329's plank crew. At 12:30 p.m. sharp, he walked off the LCI (L) 329 for the last time. Since his ship was first commissioned for duty in November 1942, he had spent the last year of his life onboard that "floating bedpan" with men he considered brothers. They crossed the Pacific and underwent their baptism of fire at Rendova together. They had exercised at General Quarters and ate all their meals together every day. They had shared stories of family, friends, and girlfriends back home, as they lay in their bunks together at night before falling asleep to the noisy hum of the diesel engines. He would never see any of them again.

Stephen reported aboard his new ship, USS LCI (L) 65, for permanent duty. He would essentially have to start from scratch. He would no longer recognize a single face aboard his new home. He would have to form new bonds with new shipmates. One of those bonds just happened to be with Stephen's first new friend, the cook of the LCI 65.

Lester Eugene Aiston, like Stephen, was also from Michigan. Aiston was a tall, husky, redheaded young man who had been aboard the 65 since July 1943. Stephen would spend the rest of his years speaking fondly of his good buddy Aiston, who spent most of his time in the galley preparing the crew's chow. One thing Stephen made clear was that Aiston was a good kid.

Stephen also hoped his new commanding officer would be better than Illing, who he never fancied much. He was relieved to find that he liked his new commanding officer, Lieutenant Robert F. Ruben, a young man from Beverly Hills, quite a bit more than his last skipper.

On January 17, his first morning aboard the LCI 65, Stephen awoke to welders and workmen coming aboard his new ship. Like the

329, the LCI 65 was also at Carter City, Florida Island, undergoing repairs and installations of new equipment. The 65 installed a compass, welded some stanchions, and repaired the radio. They would spend the remainder of January in the Govana Outlet of Florida Island preparing for their next major invasion in the Pacific.

However, fate would not be kind to Stephen's wishes, as his relief was only temporary. Robert Ruben would not be the captain of the 65 for very much longer. Although it was impossible to know at the time, Ruben would be replaced in exactly two months by the LCI 65's current executive officer, Ensign Charles J. Macaluso from Cleveland Heights, Ohio. For reasons unknown, Macaluso would offer only hostility to Stephen during his time as skipper of the LCI 65. And for the next year, Stephen would offer only resentment in return for Macaluso's actions—resentment that would last a lifetime.

————

That same month back in the States, the workers at George Lawley & Sons Shipbuilding Corporation had been laboring away in Neponset, Massachusetts, on a new design for the LCI. Launched in the first week of January, the LCI (L) 402 was the third different version of an LCI design since the war started. The first version, the LCI 1-350-class, built with a square-shaped conning tower, was how both Stephen's LCIs were designed. The improved second version, the LCI 351-class that began production with the LCI (L) 351, had a round, cylindrical conning tower and larger deckhouse. But this third version was truly the ultimate LCI. This LCI was to be a combination of the first two versions. It was built with a troop-delivering bow ramp instead of two side ramps and kept its larger deckhouse design with round conning tower. Which meant it was armed like a heavy gunboat but could still land troops. The U.S. Navy had found a way to have its cake and eat it too. This design that began with the LCI (L) 402 proved so successful that beginning with the LCI (L) 691 it became the standard design for all remaining LCIs manufactured for the duration of the war.

A side-by-side comparison: USS LCI (L) 402 with its bow ramp and USS LCI (L) 354 with its side ramps, date and location unknown. (U.S. Navy Department photo, *Allied Landing Craft of World War Two*)

8

FEBRUARY 1944: THE GREEN ISLANDS

"As a preliminary to the coming drive, I directed the seizure of the Green Islands, north of Bougainville. On February 15, Allied troops carried out a weakly opposed amphibious landing and captured the islands to cut off an estimated 25,000 enemy troops to the south. . . . For all strategic military purposes, this completed the campaign for the Solomon Islands."

– General Douglas MacArthur[1]

Stephen and the crew of the LCI 65 began the morning of February 1 anchored in Govana Outlet, Florida Island. It was a big day for the crew. At 7:28 a.m. the 65 began getting towed backwards (sternway first) by two small landing craft into dry dock to undergo repairs on her port rudder post, as well as further welding on her hull. At 7:42 the 65 pumped out her ballast tanks and hooked up the electricity through the drydock. Within minutes, welders and repairmen began installing temporary strut bearings and working on the 65's rudder post and hull. Ruben's captain's quarters also had work done on it by carpenters. After a full day's work, the carpenters went ashore, the welders and repairmen completed installation of the rudder post and welding of the hull, and two men from the 65 who had been selected to go ashore and pick up the rudder contact box from the LCI 70 arrived back to the 65 shortly before 7 p.m. with the box in hand.

At 9:45 a.m. the next morning, February 2, Ensign Cadwallader, the repair base officer, came aboard the 65 inquiring about work being done on the fire and bilge pump system in the ship's after-steering station. After all issues were cleared up, Cadwallader departed and went back ashore Florida Island. In the early afternoon, a working party from the 65 went ashore to procure pyrotechnic ammunition (tracer bullets) and returned later that evening carrying three cases. While the working party acquired their ammo ashore, Cadwallader sent several base repair workmen out to the 65 to work on the fire and bilge systems. Several hours after the workmen had finished their work and departed, the 65 beached at Carter City for the night.

Over the next week, the ship would continue repairing, refueling, and resupplying in preparation for their next landing—the Green Islands. Years later, when recalling the battle, Stephen pointed to a map of the Bismarck Archipelago and said, "Right up in here is the Green Islands. We invaded that one."

The seizure of the Green Island group, located about one hundred twenty miles east of Rabaul and forty miles northwest of Bougainville, was needed to provide a new Allied base for attacks on Japanese forces that went well beyond the previous range of aircraft in the South Pacific at that time. The base was to provide a staging area for the operation of fighter and bomber planes as well as PT boats. The Green Islands consist of four flat and heavily wooded islands, which almost encircle a lagoon. Nissan Island, where the Allied base was to be built, is horseshoe-shaped and the largest of the four islands. The atoll spans about nine miles long and five miles wide.[2]

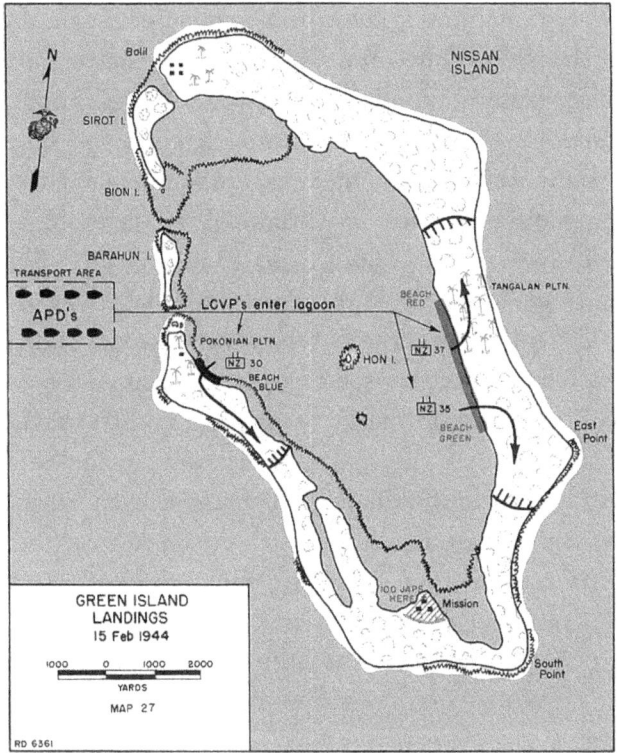

A U.S. Marine Corps map of the initial invasion of Nissan Island on February 15, 1944. Stephen's LCI (L) 65 landed New Zealand troops at Blue Beach on February 20. From *Bougainville and the Northern Solomons* (1946), by John N. Rentz, p. 117.

On February 8, the LCI 65's first African American crewmember, Steward's Mate 2/c Shellie King Sr., reported aboard for permanent duty. A steward's mate, author Mitch Weiss explained, "was the lowest rank on the ship, a position reserved for black men." It was February 1944, but at this time according to Weiss, "the U.S. military was as segregated as the Jim Crow South." Throughout every branch of the armed services, African Americans held menial jobs. As a Black man, King's shifts mainly occurred in the officers' wardroom. He was treated as a servant, as his duties included serving officers their dinners, making their beds, doing their laundry, washing their dishes, cleaning the officer's dining room for the next

day, and even remaining at their beck and call—in case the officers
needed a shoeshine. The crew's bunks in the bowels of the ship
were already brutally hot, but the steward's mates' bunks were, as
Weiss described, "worst of all, tucked against the engine-room
wall."[3] Two months later, another steward's mate named Willi Ladd
would also join the 65's crew in April. Stephen remembered, "There
were two Blacks from the South," as King was from Florida and
Ladd was from Texas. In WWII, the steward's mates were often
forced to disguise their true selves which were unseen by most
white men. On rare occasions, officers who censored the crew's mail
like Louis Harlan aboard the LCI 555, did get a first-hand look into
their steward's mate's perspectives. Harlan remembered, "C.R.
Richardson . . . was a middle-aged black man from Virginia. . . . A
lifetime of segregation had taught Richardson to wear the mask, to
dissemble in the presence of white people, to keep his feelings to
himself."[4] In a letter to his girlfriend, Russell Tye, another officer
from the LCI 555, once described the almost sadistic treatment of
Black men across all branches after he'd witnessed a certain incident ashore:

> One thing it is perhaps just as well that the States do not see . . . is
> the condition of the Negro Seabees and soldiers. Their quarters are
> a disgrace to the Nation. It's conceivable that under some conditions,
> such unsatisfactory quarters must be used, but the comparison is
> the bad part. There is nearly as much difference in quarters for the
> Negroes and whites as there is in Negro and white homes in Amer-
> ica. The Negroes have the muddy places and their tents are so
> poorly constructed that the only important difference between them
> and foxholes is that the tents keep most of the rain out. The camp
> areas look as if they were picked as the worst place possible.[5]

Each and every sailor had his own internal struggles during the
war, but one could reasonably argue that it was Black men subjected
to this inherent discrimination who faced the most difficult challenges of all.

———

On February 9, the LCI 65 started the morning beached in Berth 6 at Halavo Peninsula, Florida Island, in company with various other landing crafts and LSTs of Flotilla Five. At 8:15 a.m., base painters came aboard and began to camouflage the ship. When the painters were done, the 65 looked like a giant action figure with its various blended pattern of green, brown, and black. The camouflage would allow the ship to blend in to the nearby shore and take cover during potential enemy air raids.

Everything aboard the 65 seemed all ready to go, until they hit a setback on February 11, and the magazine started to take on water in the bilges—the area between frames at each side of the floors where seven inches of water seeped. But that wasn't all. In the early afternoon, the rudder—the vertical blade at the 65's stern used to change the ship's direction—went out. Then, to top it all off, the engine motor cut out while the ship was underway, so manual hand steering had to be used to get the ship back to Carter City to undergo repairs. The 65 ended the day beached in Berth 6, Halavo Peninsula, Florida Island, in company with various units of LCIs of Flotilla Five. They would remain there undergoing final repairs and gathering all needed provisions for the next four days.

On the evening of February 15, while Stephen and the rest of the salty crew of the 65 lounged around the ship after taking on fresh water at Hutchison Creek, Lieutenant Ruben stepped down its ramps and headed for an important meeting held by LCI Flotilla Five Commander Alfred Jannotta that would take place further past the beach. There, Jannotta informed Ruben and the men that the invasion of the Green Islands had begun. Elements of the 3rd New Zealand Division had landed on the tiny atoll of Nissan Island. Their 35th Battalion had landed on Green Beach, the 37th Battalion on Red Beach, and the 30th Battalion on Blue Beach. Jannotta informed Ruben that he would be taking the LCI 65 onto Blue Beach to help the 30th Battalion mop up any Japanese resistance with the rest of the 3rd Division. What Stephen, the crew of the 65, and perhaps even

Ruben didn't know was that the small Japanese garrison that was on Nissan Island was putting up fierce resistance.

But more so than anything, the Allies needed to secure the Green Islands to cut off any chance the Japanese could use Nissan Island to reinforce their fighters on New Britain. Stephen described it years later when pointing to the islands off eastern New Guinea on a map, saying, "We had to . . . isolate the Japs around here," as he motioned to New Ireland and New Britain Islands. He then added, "This was the tough area, this—New Britain—was where the Jap fleet was harbored in." Nevertheless, the Allied noose around the throat of Rabaul was tightening.

With this new information, Ruben departed and returned to his ship at Berth 7 at 9:58 p.m. The next morning, Stephen and the crew of the 65 departed for Guadalcanal, where they picked up the convoy consisting of LSTs and submarine chasers at Kukum Beach. The convoy departed Guadalcanal at 10 a.m. and set sail for Vella Lavella where they practiced gunnery drills for the next hour and a half. Shortly after noon, the entire convoy fired at the aerial target sleeve being pulled across the sky by a small Allied plane. Though several ships broke off independently and left the convoy due to other orders, Stephen and the convoy nevertheless steamed ahead to Vella Lavella. The last entry in the 65's log at 8 p.m. stated that the ship was just to the southeast of the New Georgia Islands group.

The convoy finally arrived on the morning of February 17, at 11:55 a.m. While Stephen's ship prepared to beach at Juno River Beach, Vella Lavella, U.S. carrier-based planes from Admiral Nimitz's Central Pacific Force had begun smashing another Japanese naval base—this one at Truk in the Caroline Islands. Having the base at Truk neutralized would help cover Flotilla Five's landing at Nissan Island scheduled for February 20.

The 65 beached at Juno River Beach at 1:19 p.m. alongside the LST 268, where Stephen and the crew would remain until the next morning. Starting at 9:00 a.m. sharp on February 18, the 65 quickly began embarking troops from the New Zealand 3rd Division. Twenty-five minutes later, the 65 was fully loaded with one hundred sixty-seven men, four officers, and ten tons of cargo from the Kiwis' 17th Regiment, 14th Brigade. Shortly after noon, the LCI 65 retracted from the beach and formed in a column with LST 118, LST 269, and the two destroyers, USS *Baron* (DE-166) and USS *Guest* (DD-472). The column rendezvoused at Point "Pencil" where they linked up with the rest of the convoy and began their two-day journey to invasion.

In the early morning of February 20, Stephen and the 65's crew spotted Nissan Island dead ahead, due north, as they were sailing off the northwest coast of Bougainville. At 6:15 a.m., six APDs (high-speed transports) and their LST escort arrived. The LSTs formed in a single column and proceeded toward the entrance of the lagoon. The 65 and the rest of the LCIs followed. At 8:22 a.m., Stephen and the crew of the 65 beached at Blue Beach 8 and lowered the ramps for the very last time as a landing-type LCI. Stephen watched from within the square conning tower for the next thirty minutes as the men of the New Zealand 3rd Division stormed down the 65's two ramps with their equipment. At 8:52 a.m., the 65 retracted from Blue Beach 8 and remained in the lagoon, on alert, in case the LSTs that were still unloading on the beach needed assistance retracting. By 3:47 p.m., all the LSTs were done unloading and had rejoined the 65's convoy. All ships formed up and headed back to Guadalcanal. The invasion of Nissan Island was complete, and the Allied forces would secure the Green Islands over the next five days.

As Stephen and the 65 headed back to the southern Solomon Islands where they would remain until the end of February, U.S. carrier and land-based planes launched the last of the bombing raids that finally knocked out Rabaul, taking it out of commission for good. Operation Cartwheel was almost complete. Or as Stephen put it sixty-seven years later, "We had 'em completely surrounded."

On February 28, 1944, Stephen's ship joined a convoy headed south to the New Hebrides Islands, where they would have a very important March. For the next month, the 65 would undergo its first major conversion into an LCI gunboat.

9

MARCH 1944: AN ODE TO A GUNBOAT

And when we take the troops aboard
To land on some enemy shore,
They take one look and pray to the Lord
They won't see us any more.

March 1944 brought the LCI (L) 65 radical change. Like so many others for Stephen during the war, he was on the move as yet another month began. Some may say the ship's most sweeping change was the major modifications the 65 underwent in Espiritu Santo, New Hebrides Islands, as the ship began its conversion into an LCI gunboat. Others may argue it was the significant increase in the number of new crewmen reporting aboard, which would be required to man all the new heavy machine guns being installed on the ship. But from Stephen's point of view, the most noticeable change from March was the change of commanding officers that would occur on March 17.

Espiritu Santo, the largest island of the New Hebrides group, had been a major Allied air and navy base since early 1942. Though

Stephen had sailed past the New Hebrides twice on his journey aboard the LCI (L) 329 in June and December 1943, this was his first time making a stop there. Espiritu Santo contained a myriad of depots and airfields, which made it a critical piece of the Allied war effort and supply chain. It was here in this small corner of the Pacific that the Navy's massive re-outfitting first kicked off for large-class LCIs, which were turned into LCI gunboats, rocket ships, and mortar ships. Dozens upon dozens of LCI (L) ships entered Espiritu Santo during WWII and each came out reclassified as either LCI (G), LCI (R), or LCI (M).[1] This included the LCI 65, which would later be renamed LCI (G) 65 several months down the road.

On March 2, USS LCI (L) 65 arrived at Segond Channel, a small channel on the southeastern edge of Espiritu Santo Island, between the main island and much smaller Aore Island. Once they passed through the anti-submarine nets, Stephen and his shipmates moored in Berth 1 with the LCI (L) 66 on her starboard side. One of the 65's officers, Ensign McKeon, took a working party of crewmen ashore to acquire provisions, and by 10 p.m., the ship once again had 2,800 gallons of fresh water aboard.

The real work started early the next morning shortly after 8 a.m., when the 65 proceeded alongside the repair ship USS *Oceanus*, where their conversion into an LCI gunboat began. Upon securing themselves along the port side of *Oceanus*, representatives from General Motors came aboard to check the gearbox. The crew would spend the coming week resupplying, repairing the guns and engines, and conducting gunnery training. As Stephen and the crew conducted their duties aboard the 65, the LCI 64 was also undergoing similar conversion along with them. On March 10, shortly after 10 a.m., a small fire caused by the welders had to be put out by the firemen aboard the 65 using the CO_2 extinguisher. By noon, the magazine sprinkler system and magazine were reported to be in normal working condition, which was good news for everyone onboard.

Throughout March and the months to follow, the 65's crew would be vastly expanded. Later that afternoon, the first of the 65's new crewmen arrived. The new guys would join the ship's expanded gun crews for her additional heavy machine guns because they were needed to operate the new equipment. But there was a consequence to these additional bodies. As LCI 74 veteran Robert Kirsch pointed out, "When the LCI(L)s were converted to Gun Ships, the crew grew to more than 60 men. Imagine how hectic it would get when 60 men were trying to use one shower stall, one washbasin, two hoppers, and one urinal all in a very small compartment called a head."[2]

The first new guy to call the LCI 65 his permanent home away from home was a radarman, Seaman 2/c Charles Bligh—whose real home was in Jerseyville, Illinois. Two days later, on March 12, thirteen additional new sailors reported aboard and were added to the 65's crew list. One of them was a rowdy Seaman 1/c from Huntingdon, West Virginia, named Henry Haymen Reeder. The other was James Kent Reid from Lynchburg, Virginia. Another, a young man from Pittsburgh, Pennsylvania, named Joseph Orbich, would eventually become the 65's second quartermaster assigned to the conning tower's pilothouse alongside Stephen. Then, on March 14, four more crewmen reported aboard, including another Gunner's Mate 3/c from Detroit, Michigan—Stuart Grant Oberson. In less than a month, one more seasoned old salt would be named to the crew named Paul DeWitte. When he joined the crew the first week of April, DeWitte was one of the oldest men aboard, as he'd enlisted in December 1940, almost a year before Pearl Harbor. Over the next several months, Stephen began to form a new group of friends with Reeder, Reid, Orbich, Oberson, DeWitte, and as previously mentioned, Lester Aiston. They'd become close friends and "partners in crime." Stephen and James Kent Reid became especially close, as he and Reid would eventually spend hundreds of hours together within the conning tower. Stephen's job as the ship's quartermaster required him to work especially closely with Reid, the signalman—as their duties called for them to be together up in the conning tower every time the ship was underway. Stephen navigated and Reid sent the signals using his semaphore flags or signal light.

Life aboard the 65 changed seismically for Stephen on March 17. After taking on fresh water in the still-dark early morning hours, welders and workmen began coming aboard from the USS *Oceanus* like any other workday. But then at 8:30 a.m., the crew was mustered to quarters where a formal announcement was made. Pursuant to orders, the ship's current commanding officer, Lt. Robert Ruben, was detached from duty and sent back to the continental U.S. His replacement was a young Ensign from Cleveland Heights, Ohio, who played football for Notre Dame named Charles Joseph Macaluso. There were many vivid memories from World War II that Stephen shared years later, but it was the visceral, detailed accounts of incidents involving Macaluso that Stephen recalled clearest and, at times, with the most emotional pain. Stephen stated on several occasions that he did not respect Macaluso for how he treated his crewmen and the decisions Macaluso made throughout the war. In time, Stephen and his buddies would experience Macaluso's wrath firsthand.

With his duties handed off to Macaluso, Ruben said his farewell to the crew and departed the ship the next morning. Several hours after Ruben's departure, three more gunners were added to the crew list, reporting aboard the 65 shortly after lunch. On March 19, thirteen of the new crewmen, including Stephen's new friends Oberson and Reid, attended gunnery school on the island, and on March 20, Ensign John F. McGinnis reported aboard for duty as the 65's new gunnery officer. Three days later, on March 23, welders and workmen from the USS *Oceanus* came aboard the 65 with six .50 caliber machine guns and proceeded to add them to the ship's new permanent arsenal of firepower. On March 25, Oberson, Reid, and the others returned to the ship from their gunnery training. Stephen and the crew spent the next four days completing routine maintenance work, refueling, and restocking ammunition. Then, on March 29, the force maintenance officer officially declared the LCI 65 as completed with its conversion. They cast off their lines and got underway to move to the shore, where they would beach and rest for the night. The next day would be a big one for Stephen and his shipmates fresh out of gunnery training.

On March 30, at 11:53 a.m., Stephen and the crew retracted from the beach and proceeded out of the harbor to conduct gunnery exer-

cises to test the new gun installations, in company with LCI 61, LCI
64, and LCI 66, all of which had also undergone conversion. First, the
crew tested their new 40mm by firing eleven rounds of both tracer
bullets and standard bullets. The gun was reported in good working
order. Next, the crew tested the new three-inch 50 caliber machine
gun, which was now their heaviest gun. They tested six "star shell"
rounds, which were special rounds used at night to aid nearby
infantry and marines fighting on land. Star shells were fired up into
the night sky, where they exploded like a firecracker with a tremendous flash. According to author Mitch Weiss, "Each flash released a
giant flare on a parachute, which drifted slowly back to earth,
swinging like a silly afterthought."[3] The three-inch 50 caliber
machine gun was reported in good working order. A .50 caliber
machine gun was also tested fired and reported in good working
order. The LCI 65 was now ready to take its new machine gun armament into battle. Over the next year, this new addition of machine
guns on the 65, as well as countless other LCIs, would be a critical
component of Pacific invasions to come. When the LCI (L) 65 was first
commissioned on December 14, 1942, the only guns it had aboard
were four single 20mm guns and two .50 caliber M2 Browning air-
cooled machine guns. By June 1944, the 65 had aboard the following
guns:

One (1) 3-inch 50 caliber Mk. 22 dual-purpose heavy gun
• Mounted at the well deck in front of the conning tower near the
front of the ship. This was a heavy and powerful all-purpose gun. It
could fire at targets on land, air, and sea. The "3-inch" meant the gun
fired a projectile three inches (76mm) in diameter, and the barrel was
fifty calibers long (or about 3.8m).

One (1) 40mm Bofor automatic gun
• Mounted behind the conning tower above the deckhouse (amid
ship). This was a powerful single barrel machine gun that took a
team of seven men to operate.

Four (4) Oerlikon 20mm machine guns
• Fed by a sixty-round snail drum and manned by a team of two to five men. Two guns were mounted next to the 40mm, behind the conning tower atop the deckhouse, and two were mounted at the rear of the ship, one on the port side and one on the starboard side.

Six (6) .50 caliber M2 Browning air-cooled machine guns[4]
• Three on each side of the ship—a last resort against air attacks.

––––––

Other LCI crews were having the same experience as Stephen and his buddies. Around this time, Signalman 2/c James D. Robertson remembered first transferring aboard the gunboat LCI (G) 64. In his book he shared his experience of being aboard an LCI during this time of massive conversion writing, "Gun tubs were being installed requiring intensive cutting and welding. New guns were being brought aboard still covered in Cosmoline (a rust preventive) that had to be removed with solvent. It seemed every part of the ship was undergoing some sort of overhaul. To make matters worse we had to live aboard while all this was going on—and as a matter of fact, it went on 24 hours a day."[5] LCI 70 veteran Gil Ortiz added, "We called it the gun tub because we cut holes in it and when it rained we would plug up the holes and collect the fresh water for showering."[6] A different signalman, Eddie Benoit of the LCI (R) 74, shared a poem by an unknown author years later titled, "The LCI Blues," that humorously expressed the attitude of LCI sailors toward their ships:

You've heard of the cans and the APAs
And you've heard of the cruisers, too
And you've also heard of the fast PTs
And some of the things they do.

But, brother, if you've a minute to spare
And feel in need of a cry
Then sit right down, and my tale I'll share
Of the terrible LCI.

They're a helluva horrible looking mess,
Neither ship nor barge, it's true,
They're a joke to the fleet, I must confess
But not to the poor damn crew.

She will rock and roll on the calmest day,
She'll buck like a kangaroo
And pitch in the most peculiar way,
Though all is serene on the blue.

The engineers sit down in the miserable hole
And can't even hear or think,
While the boys topside, as they sway with the roll,
Have often wished she would sink.

The signalman clings to his light in the conn,
While she heels over fifty degrees,
And the helmsman struggles to hold her on,
Though it's blowing a very light breeze;

The cooks in the galley sob and moan
As over the stove slops the stew,
And the crew sets up a terrible groan,
And so, by God, would you.

And when we take the troops aboard
To land on some enemy shore,
They take one look and pray to the Lord
They won't see us any more.

They were built, I am told, of just old junk,
And stuck together with glue,
The whole thing was planned by a humorous drunk,
An inmate of Annapolis U.

But don't pity us, friend, and dry that tear,
Though we are thankful for your grief,
but we've sailed in this thing for nearly a year
So, please, let's have some relief.

Another fellow skivvy waiver, Signalman 3/c Winston LeRoy of the LCI (L) 564, also expressed his attitude through a poem titled "An LCI Creed" he shared in 1995:

We are the boys of the LCIs
An unsung lot we are
We battle planes of the sky
And fight against the sea
Our crew is small, the men are tough
You never see them fall when things get rough
We sail the seas day and night
Fighting for what we know is right
We lead the first wave to the beach
Under God's protective reach
You read what the battleships did
Do you think we ran and hid?
Our guns are small their fire is weak
With enemy mortars we play hide and seek
We knock down enemy planes in flame
While a battleship collects fame
Despite the enemy, rain, wind, and sea
We also wish to stay free[7]

10

APRIL 1944: BACK TO THE SOLOMONS

"Ship building is America's greatest pride, and in which she will, in time, excel the whole world."

– Thomas Paine, *Common Sense*

In the early morning hours of Sunday, April 2, the LCI 65 cast off her lines and proceeded out of the harbor. She was heavily armed and looking fresh. The boys from the paint barge had put the finishing touches on the 65 the day before. Stephen and the crew were now headed back to the Solomon Islands with a new camouflage pattern painted on their ship. They formed a double column with the LCIs 61, 64, 66, 518, and 519 after passing through the submarine nets. The two-day journey was a rough one for the 65's crew. They endured strong winds and rain that caused one of their ship's port engines to break down.

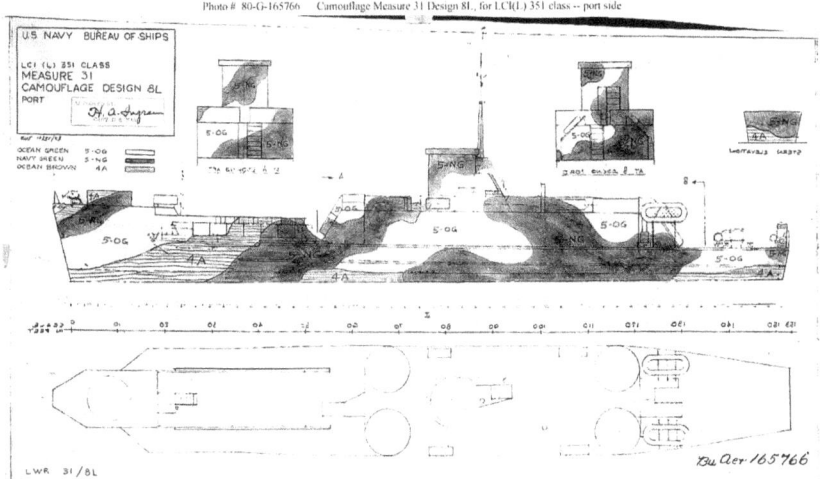

A blueprint of a round-conned LCI (L) 351-class camouflage paint design. Many variations of this paint pattern were used to help disguise ships of the Amphibious Forces during World War II. (U.S. National Archives photo No. 80-G-165766)

At exactly 9:50 a.m. on April 4, Stephen spotted Guadalcanal from inside the conning tower. As she steamed ahead within the convoy of LCIs, the 65 suddenly stopped her starboard engine at half passed noon. A piece of the ship's propeller—the strut bearing—was reported either loose or missing. This meant that after her long trip, she'd have to finish the journey getting towed into Hutchinson Creek, Florida Island, by the LCI (L) 66. Stephen and the crew remained anchored there until the next morning but were able to make it to Carter City on their own, just across the bay, where they'd enter drydock and undergo repairs over the next four days. Just before they entered in drydock on April 5, brand new crewmember, Seaman 1/c Paul DeWitte, reported aboard the 65 for permanent duty. As a fellow Midwesterner from Chicago, DeWitte and Stephen quickly became friends. DeWitte would eventually strike (train) for the duty of Fireman.

As mentioned, DeWitte, one of the oldest crewmembers, enlisted in December 1940 before the war had started. He was slightly older than Stephen and the rest of the 65's enlisted crew and was already in the Navy when Pearl Harbor was attacked. He had served aboard the

USS *St. Louis* before transferring to the 65. But age didn't stop Paul Cyriel DeWitte from having quite a knack for getting in trouble. While aboard the *St. Louis* in May 1942, DeWitte had fallen asleep while on watch and been sentenced to solitary confinement on a diet of bread and water for a month. DeWitte's record was not spotless when he came aboard Stephen's ship, and in the months that followed he would continue to add blemishes to it.

Stephen and the crew were allowed to leave the ship on Sunday morning, April 9, to attend church services at Carter City. After quickly receiving their blessings from the Lord, the men wasted no time returning to the ship shortly before 8 a.m. Sabbath or not, it was always a workday in the South Pacific. Within a half hour, the LCI 65 was once again underway, with Stephen guiding her back across the bay to Tulagi Harbor. The 65 beached at Government Wharf, Tulagi Harbor, in company with the LCI (L) 359. Soon after, a working party left the ship to procure supplies and provisions as the ship commenced taking on 6,800 gallons of fuel. Once the work party returned to the ship with the needed provisions, the chronometer was wound at noon and the fuel was secured. The LCI 65 retracted from the beach and got underway for another trip across Iron Bottom Bay, returning to Hutchinson Creek—a place that had become so familiar to Stephen and the crew.

The men lived a straining life of standing four-hour watches and being on call for the other twenty. Even as Florida Island became an increasingly smaller corner of the advancing war in the South Pacific —pushing it further from Japanese threat with each passing day— the men would of course still have to arise in the middle of the night at a moment's notice. Even harmless routine visits from Washing Machine Charlie[1] could trigger General Quarters, waking the men from their slumber and scrambling them to their battle stations in complete darkness. If this happened, which it so often did, it was the perfect recipe for bumped heads and bruised limbs. Stephen and the men were tired. They just wanted to sleep. They were always tired and hungry. And that ever-present seasickness didn't help. For a ship the size of an LCI, there was always the seasickness. That, too, had become a familiarity to the men in this war. And for the weary young men aboard the 65 who did manage to sleep, this night in Berth 4 was no exception.

The 65's new captain, Lt. (jg) Macaluso, mustered the crew at 8 a.m. the next morning, April 10. He and the executive officer, Lt. (jg) Philip Sadtler, gave instructions on various gunnery drills that were to be conducted in the days to come. This included showing off its new firepower to Navy brass. Lieutenant Commander E. P. Rankin was coming aboard to watch the gunboat's new machine guns in action. Macaluso didn't want any mistakes. Starting at 8 a.m. on April 12, Rankin reported aboard the LCI 65 for four days duty. Joining him were an Ensign, a Yeoman, and a Steward's Mate. Several hours later when the 65 arrived at Rua Sura Island off northeast Guadalcanal, the gunnery drills began. A working party was sent over to set up targets on land. Shortly after 5 p.m., the number two .50 caliber machine gun was test fired, and after 175 rounds, Rankin had seen enough for the day. The first test of the first gun of their new gunboat armament was a success.

The next morning, Rankin, the men accompanying him, and the crew of the 65 hoisted the anchor and got underway for gunnery practice once again. This time the 65's crew tested the biggest gun of them all—the three-inch 50 caliber. In naval gun terminology, the "three-inch" meant the gun fired a projectile three inches (76mm in diameter, and the barrel was fifty calibers long (about 3.8 meters). Even six decades later, Stephen never forgot about "the biggie" as he described it. "We was heavily loaded for a little fella," he recalled. The men from the ship watched as the large three-inch shells ripped through their targets. Even as teenagers, the young men were aware of its destructive firepower. The powerful gun would become a critical aspect of patrol missions and invasions to come for the small LCI. Soon, the LCIs 23, 64, and 66 joined the 65 for gunnery practice. At around 10:15 a.m., Lt. Commander Rankin left the ship with gunnery officer, John McGinnis, to inspect the targets. They both returned to the 65 several hours later to report that all targets had been destroyed.

On April 14, the 65 returned to Hutchinson Creek, Florida Island,

and beached in Berth 3 after a routine day. Shortly before sunset, Lt. Commander Rankin, Ensign R. B. Cleland, Yeoman 2/c R. P. Peterson, and Steward's Mate M. A. Mueller departed the 65 for the last time. Stephen would never see Lt. Commander Rankin again. Two days later on April 16, Steward's Mate 3/c Willi Ladd, the second African American of the 65's crew reported aboard, along with Seaman 2/c Albert Mahan, a baker.

After spending several days around the Govana Inlet practicing tactical maneuvers and losing an anchor in the process due to a defective shackle, it was finally time to test the 65's heavy artillery during a practice landing. In the early morning hours of April 18, the 65, as well as other LCIs from Flotilla Five, formed into their formations and entered the Sandfly Passage off the northwestern coast of Florida Island. The 65 waited until just after sunset when they and several other LCIs began their landing operations. The 65 and several LCI gunboats pummeled their targets on the flanks of the landing beaches, as a few other LCIs landed and simulated securing and establishing a beachhead. The 65 only needed to use one round from the 3-inch 50 caliber gun. It completely obliterated the target.

Stephen and the crew of the 65 spent the remainder of April around Hutchinson Creek and Carter City repairing, painting, refueling, and resupplying their ship for the coming weeks. However, it should be noted that on April 27, Seaman 1/c Walter Winfred Henry Jr. reported aboard the 65 with baggage and papers, being recently transferred from Flotilla Five. Henry was born ten days before Stephen on August 15, 1924. Henry was from Maryland Heights, Missouri, and had enlisted in the U.S. Navy on June 8, 1943. Henry, a handsome young man with dark hair, would serve on the 65 for the next four and a half months until a gruesome hand injury inflicted at Morotai took him out of the war for good.

11

MAY 1944: THE BROKEN BISMARCK BARRIER

"The real heroes of the landing were the rocket-firing LCIs. They steamed in close to the beach, remained on the flanks of succeeding waves of assault boats, and poured in deadly cover of rocket fire."

– Admiral Daniel E. Barbey[1]

Stephen and the crew of the LCI 65 began this month beached in Berth 7, Hutchinson Creek, with various other landing craft. They would spend the first week of May there undergoing repairs and taking on supplies.

On May 5, the Office of the Commander of the Third Amphibious Force sent out a letter to all commanding officers in the South Pacific. The letter was written by Admiral William "Bull" Halsey, recognizing all men in the South Pacific armed forces for their contribution in the South Pacific campaign. The successful landing of the 4th Marine Division on Emirau Island in March 1944 marked the last invasion in the series of operations meant to encircle Rabaul. His letter was a

declaration of Allied victory in "Operation Cartwheel" and in it
Halsey praised the men, writing:

> With the announcement of the virtual completion of the South
> Pacific campaign, except for mopping up and starving out operation,
> I can tell you and tell the world that no greater fighting team has
> ever been put together. From the desperate days of Guadalcanal to
> the smooth steamrolling of Bougainville and the easy seizure of
> Green and Emirau, all U.S. and Allied services put aside every
> consideration but the one goal of wiping out Japs. As you progressed
> your techniques and team work improved until at the last ground
> Amphibious, Sea, and Air forces were working as one beautiful
> piece of precision machinery that crushed and baffled our hated
> enemy in every encounter. Your resourcefulness, tireless ingenuity,
> cooperation, and indomitable fighting spirit form a battle pattern
> that will everywhere be an inspiration. And a great deal of the credit
> for the sky-blazing, sea sweeping, jungle smashing of the combat
> forces goes to the construction gangs and service organizations that
> bulldozed bases out of the jungle and brought up the beans and the
> bullets and supplies. You never stopped moving forward and the Jap
> never could get set to launch a sustained counter attack. You beat
> them wherever you found them and you never stopped looking for
> them and tearing into them. Well done.[2]

However, despite Halsey declaring victory, the 65's job was far
from over in the South Pacific. In fact, Stephen and the crew were
using the first week of May to prepare for a month-long patrol
mission "mopping up and starving" Emirau Island that Halsey spoke
of in his letter. The weary young men would soon be thrust back into
action.

On Sunday, May 7, Stephen and the crew left the 65 to attend
church services on Florida Island. It would be the last time they'd be
on dry land for more than a month. Early the next morning, May 8,
the crew of the 65 set out for Guadalcanal just across Iron Bottom
Sound. They waited for several hours until the LCI 64 had finished

taking on fifty troops and Navy personnel, and then departed for Vella Lavella in a convoy with LCIs 327, 223, 64, and 329—Stephen's old ship.

The convoy of LCIs arrived at Barokoma, Vella Lavella on the morning of May 9. The 65 once again anchored out to sea and waited for the other LCIs to unload their troops. Later that afternoon, the 65 departed for Bougainville in company with the LCI 64 and PC 1138. The next morning, the three ships made a stop off at Torokina Point, Bougainville, then departed for their three-day journey to their final destination: Emirau Island.

The LCIs 65, 64, and PC 1138 arrived at Homestead Lagoon, Emirau Island in the early morning hours of May 13. Emirau Island sits almost right on the equator with Homestead Lagoon located on the island's western coast, where the island concaves inward in shape. Upon their arrival, Macaluso left the ship to get further orders and then returned shortly afterward with the 65's next assignment. Stephen and the crew were to conduct their first of many patrols off their next target: New Ireland Island.

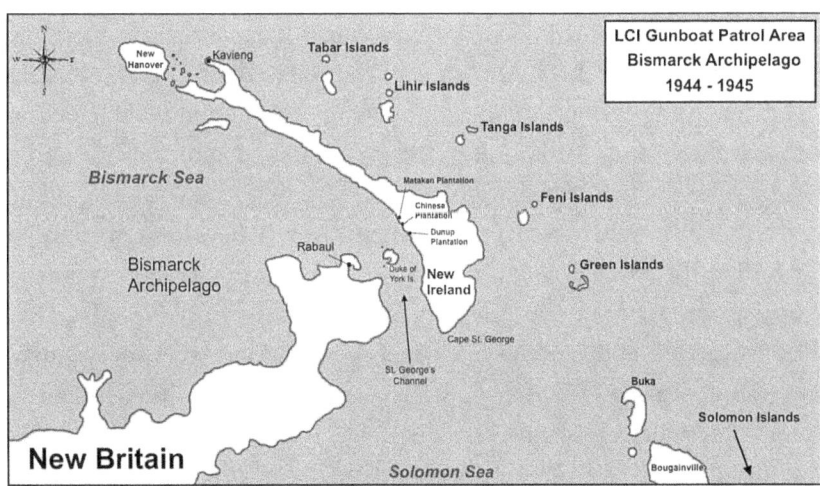

New Ireland Island, a wide but narrow island that wraps around the northeast coast of New Britain, sits right across the channel from Rabaul. New Ireland had been the target of heavy Allied bombardment since February but it was still occupied by Japanese attempting to resupply and reinforce Rabaul. As Stephen explained it years later while looking at a map, "We'd go on patrol in this area here," he said as he motioned between New Britain and New Ireland Islands. "That's where we did our patrol, because the Japs used to go back and forth across the islands. So we just put a plug on that."

But New Ireland Island also presented a unique threat. It was full of plantations scattered all over the long, narrow island. It was explained to Stephen and the 65's crew that their objectives were to destroy any Japanese plantations they found close to the coast they patrolled. However, what Stephen and the rest of the crew didn't know at the time was most of the plantations were run by Korean laborers who were captured by the Japanese and forced against their will to work in the Japanese war machine on New Ireland Island. It was common practice to work the Korean laborers—who the Japanese thought to be racially inferior—to death.

At around 9:30 p.m. on May 14, the 65 sighted several flares in the distance off their starboard side toward New Ireland Island. A little over an hour later they saw several more flares in the vicinity of Lebrechtshof Plantation. Almost as if that were their queue to begin, the 65 got underway on a special patrol mission off the north coast of But-But Bay and Lussok Bay New Ireland, looking for enemy Japanese in the dark of the night.

The 65 spotted two PTs off the starboard bow that were also patrolling the coast. Several hours later, the 65 had finished its patrol. No Japanese or plantation activity was observed from New Ireland. The 65 turned and headed back to Emirau Island with their patrol mission completed for the night. The most eventful thing from the mission actually occurred early the next morning on their way back. Stephen was in the pilothouse navigating the ship when suddenly the ship lurched violently. What was certainly believed to be a torpedo turned out to be a school of three large black fish that raked over one of the ship's propellers. Crisis averted. The 65 made it safely back to Homestead Lagoon, Emirau Island at 10:30 a.m. on the morning of the 15th. The men were exhausted and weary from an all-night patrol.

For the lucky men not on watch, they did their best to get some rest, as the LCI 65 remained anchored in the lagoon until the following day. In the early afternoon, Macaluso went ashore Emirau Island once again to get the orders for the ship's next mission. The instruction Macaluso received was to patrol off Hamburg Bay, located on the northern end of Emirau Island. At 9:15 a.m. on the morning of May 17, Stephen and the crew of the 65 got underway to relieve the LCI 64 and take over their patrol mission off Cape Ballin, the furthest point north on the island. By the time the men arrived shortly after 10:16 a.m. the seas were calm. After completing patrol sweeps of the area at various courses and speeds, Stephen and the crew of the 65 departed from the station. The patrol was negative, and no enemies were spotted. By 2 p.m. the 65 was back in Homestead Lagoon to prepare for their next patrol. The next evening, the 65 was back patrolling the same area off Cape Ballin; the seas were still calm. But this patrol lasted all night.

After spending a dark night completing seven sweeps patrolling northern Emirau Island, Stephen's ship received different orders on the morning of May 19 from the Commander of Motor Torpedo Boat Base 16. The 65 was to proceed two miles north of Mussau Island to locate and tow two PT boats back to the base, as they had run out of fuel. Stephen watched from within the pilothouse as a PBY plane from above aided the 65 in locating the distressed PTs, and by 11:56 a.m., the signalman atop the conning tower spotted the PTs. But when the 65 came alongside to lend assistance, the crews of the PTs resisted any help, believing that they could still return to base under their own power. Hence, the 65 proceeded back to Homestead Lagoon where they arrived at 4:40 p.m., their duties finally over for the day.

The next day, May 20, the 65's crew had gunnery practice with the LCI 64. They fired at floating targets six miles northwest of Cape Ballin, Emirau Island, until 2:59 p.m., at which point they secured all guns and departed back to Homestead Lagoon. But instead of resting for the night, the 65 received orders an hour later informing them they were to head back to where they had just come and patrol the northern end of Emirau Island. When the crew of the 65 arrived three miles northeast of Cape Ballin, the seas were still calm. Again, the men patrolled all night. Six and a half sweeps this time. Patrols were negative, no Japanese sighted. The 65 arrived back in Homestead Lagoon shortly before 8 a.m. on May 21. A little over six hours later, the 65 was pulled back into patrol duties, this time to their south, off New Ireland Island near Lussok Bay. By the time the 65 arrived off the coast of New Ireland, it was the middle of the night. The 65 recorded arriving off Ululnono Plantation about thirteen miles below Cape Sass. These patrols were in dangerous waters. Stephen and the rest of the sleep deprived crew would patrol them all night, in choppy seas.

After another negative patrol with no enemies sighted, the 65 arrived back to Homestead Lagoon at 11:47 a.m., where they remained until the next evening. Stephen's ship would conduct numerous patrols in this same fashion until May 29. Upon returning to Homestead Lagoon, Stephen and the rest of the dead-tired crew, who at this point were struggling to stay fully alert, received some most welcome news: they had the next several days to rest. The last three days of May 1944, the crew of the 65 remained anchored in Homestead Lagoon getting some much-needed rest for the coming summer.

12

JUNE 1944: CRIME AND PUNISHMENT

"Every time you got a chance you would get drunk."

– Royal Wetzel, LCI (G) 70[1]

June 1944 turned out to be a monumental month for the Allies in World War II. In the Central Pacific Area, Admiral Chester Nimitz was preparing the Allies to take the fight to Saipan in the Mariana Islands. Saipan would turn out to be one of history's most hideous battles. Nimitz's invasion fleet, transporting the expeditionary forces for the invasion of Saipan scheduled for June 15, 1944, departed Pearl Harbor on June 5. With the capture of the Mariana Islands, the Allies could finally have a base to unleash on Tokyo their new land-based dreadnaught bomber—the B-29 Superfortress. If the Allies could capture the Northern Mariana Islands, they would finally be in range to bomb mainland Japan, which meant one step closer to victory. In the words of Ernie Pyle, "The Marianas happen to be a sort of crossroads in the Western Pacific. . . . Whoever sits in the Marianas can have his finger on the whole wide web of the war."[2]

———

At the exact same time on the other side of the globe, General Dwight D. Eisenhower was preparing the entire Allied Expeditionary Force to land at Normandy, France, opening a western front against Nazi Germany. Known as Operation Overlord, or D-Day, American, British, and Canadian forces were originally scheduled to land on the beaches of Normandy on June 5, but bad weather caused General Eisenhower to delay the invasion by one day. Instead the armada began crossing the English Channel the night of June 5.

One of those Americans involved in the coming invasion was a good-looking young man named John P. Cummer. Seaman 1/c Cummer was a gunner's mate aboard the LCI (L) 502. As part of LCI Group Thirty-One and the D-Day invasion force, the LCI 502 had departed England the night of June 5 loaded with one hundred ninety-six men and officers of the British 8th Army headed for Gold Beach. Cummer and his buddy, Seaman 1/c William F. Miller, Jr., stood the bow watch from 8 p.m. to midnight as the LCI 502 crossed the English Channel. Cummer, a gifted writer, described in his personal memoirs what he did aboard his LCI the night before the invasion:

> Jolly Miller and I shared the bow watch from [8 PM] to midnight that night. . . . when we were relieved at midnight, I felt the need to be alone. . . . Finding a place to be alone on a 153 foot landing craft crowded with 196 troops can be a problem, but I had my own private place, cramped though it was. Under the fantail deck was the small magazine where ammunition was stored. As the Gunner's Mate, I had the key to that small cubbyhole and so it was to that place that I retreated for my quiet time. . . . I sat on the cold, steel deck, surrounded by the cases of ammunition and read from the New Testament . . . The guide to references inside the front cover had suggestions for special times. One was "for times of peril or danger," and it directed me to *Psalm 91*:

He that dwelleth in the secret place of the Most High shall abide under the shadow of the Almighty. I will say of the LORD, He is my refuge and my fortress: my God, in Him will I trust. Surely he shall deliver thee from the snare of the fowler and from the noisome pestilence. He shall cover thee with his feathers, and under his wings shalt thou trust. Thou shalt not be afraid for the terror by night; nor for the arrow that flieth by day; nor for the pestilence that walketh in darkness; nor for the destruction that wasteth at noonday. A thousand shall fall at thy side, and ten thousand at thy right hand; but it shall not come nigh thee...

Cummer stared at that last line:

... but it shall not come nigh thee ...

John Phillip Cummer, who served aboard USS LCI (L) 502
during World War II. (Courtesy John Cummer)

Hidden within his dark, private hiding place underneath the back of the ship, Cummer remembered, "A deep sense, not of fearless bravado, but of assurance in the protection of a sovereign God came to me as I read those verses." Every time he reread them, he wrote, "that tiny steel cubicle, surrounded by cases of ammunition, pitching with the motion of the sea, comes immediately to mind." He then gratefully added, "In God's providence, that protection was afforded to me and my shipmates on D-Day, June 6, 1944."

As part of the largest amphibious invasion the world has ever seen, Cummer and the LCI 502 landed the British at Gold Beach hours later. Nearby, the U.S. Eleventh Amphibious Force began landing American forces at Utah and Omaha Beaches. As British forces landed at Gold and Sword Beaches, the Canadian forces landed at Juno Beach shortly after sunrise. The ships of the U.S. Amphibious Forces landed massive numbers of men, tanks, and supplies under intense, hellish Nazi fire. The Gator Navy was arguably the most important element to the Allies' ability to eventually recapture Europe and defeat Hitler.

———

As America waged total war in Europe and the Central Pacific in the month of June 1944, it would also be a month to remember for Stephen and the crew of the LCI 65 in the South Pacific. They were still anchored in Homestead Lagoon, Emirau Island, on Monday, June 5. It was their last "day off" until they had to conduct patrol duties beginning again later that evening. Having a routine day off, in which the crew hadn't received any orders, didn't really help the crew catch up on sleep. The flat bottom of the LCI ensured that even in calm seas, the crew felt every wave as they lay in their bunks, just inches away from the bunk hanging above them. If the waves didn't cause a sailor to bump his head, sometimes they'd simply throw a man from his bunk all together. Needless to say, catching up on hours of much needed sleep was mostly out of the question. But the men did their best and got what rest they could.

Horace R. Turner, who served aboard LCI (G) 66 at this time, remembered these endless patrols saying, "We went on patrol duty the next six months in St. George's Channel where we blew up (12) floating mines and participated in shore fire just about every night.

Our schedule was 10 days patrol and 3 days rest."[3] Charles Ports of the LCI 23 wrote, "We were then assigned to patrol the St. George Channel between New Britain and New Ireland islands to stop any resupply effort between the Japanese who occupied that area. We were there three days in and one day out for rest and then back in. While in the Rabaul area, the most eerie sound would be heard at night when the area was being bombed. The bombs would scream and sound like they were coming down on you. The Japanese must have been terrified."[4]

Royal Wetzel, the cook aboard the LCI 70 also remembered this duty saying, "We patrolled St. George's Channel" and reiterated that during these months "we did a lot of patrolling." He described how "the Japs were supposed to be taking stuff from one island to another island across the channel at night. We would go in there real slow. We were at General Quarters all night. There were barges going across. We would shoot these barges up. When it got daylight, there was one [barge] we pulled alongside. A couple of guys went aboard this barge," he said, adding they found "Henry Ford wrenches on there."[5] But this patrol duty was extremely dangerous work. In his interview, Wetzel shared a tragic story relating to one of his shipmates from his hometown named Eugene Whalen that got a hold of a Japanese flare:

> This one guy got a Japanese flare, and I don't know how long he had that. He got it out and he was going to take it apart [to] get the parachutes out of it. When he did, I was on one side of him and another fellow, a gunner's mate, was on the other side of him, and he was prying it apart. There were a lot of tubes in it, little tubes. He got one of these tubes out and he got his pocketknife out and he tried to pry the end off and it blew up on him and blew his guts out. There was meat all over the side of the ship. I had hold of him and he asked me, "Am I going to die?" By that time he was gone. Then we took him over to the island and we dug a hole and put him in there. We couldn't bury him at sea, so we dug a hole. As fast as we were digging this hole, water was coming in. We just took him and put him in and covered him up.[6]

At 6:16 p.m., the LCI 65 departed Homestead Lagoon for yet another all-night patrol off Cape Tietgens, Emirau Island. Stephen and the crew were finished with their patrol and anchored back in Homestead Lagoon by shortly after 7 a.m. However, something unknown occurred at 3:20 p.m. that caused a fire in the crew's quarters below deck. It was severe enough for the 65 to send out a distress signal to call for a fire rescue party. Yet, within five minutes, the fire was extinguished with no injuries recorded. As the 65's crew was scrambling to put out the fire in the crew's quarters in the South Pacific, Eisenhower's invasion of Normandy was beginning at the same moment at 6:30 a.m. local time in France on the other side of the world.

Almost immediately after the fire was extinguished, Macaluso ordered the 65 on another patrol mission. Though the fire was surely unexpected, the LCI 65 had already gotten orders for their next mission that was expected to take place at the same time the fire was put out. Macaluso had received secret orders to conduct a special patrol mission to their southeast, just off New Ireland Island. The Allies had received secret intelligence that the Japanese still had a functioning ammunition plantation they were using against the Americans to arm their small pockets of resistance. The LCI 65, with the help of the LCI 64, were tasked with destroying this Katu Plantation, which would be in range of the LCI gunboats if they could get close enough to the coast of New Ireland. At 4:31 p.m. on the evening of June 6, Stephen and the crew got underway for New Ireland Island. Within hours a pitch-black night fell over the Pacific as they steamed toward New Ireland. At 2:25 a.m., General Quarters was sounded and the ship's crew went to their battle stations. Stephen would be part of the action from his usual place within the pilothouse of the conning tower. One hour later, at 3:27 a.m. local time, the 65 arrived two miles off Katu Plantation. Six minutes later, the 65 crucified the darkness as she opened up and began shelling the plantation with her three-inch 50 caliber and 40mm guns. Stephen remembered the exact moment with vivid clarity years later, saying, "This one night, we went on patrol. We went all the way up this one river, darker than hell, and when we got in position, that's all I could see was the silhouette of a big warehouse. So we turned the ship around, everybody got in position, and we opened up." The 65 quickly lit up the shore as the plantation began to explode into eruptions of fire. The young men made quick work of the Katu Plantation, rejuvenated by their dramatic and quick success. Clearly, the confident men thought, there couldn't have been any survivors after that bombardment.

But as Stephen and the crew later found out, Katu Plantation wasn't all that it appeared to be. It turns out the Americans were slightly inaccurate in their original assessment of the plantation. It was indeed a functioning plantation that was being used by the Japanese, but what the Americans didn't know at the time was that the plantation was being operated entirely by aforementioned Korean laborers who were forced against their will to work for the Japanese war effort. Stephen continued, "We killed 180 Korean laborers. We thought they was Jap soldiers." When asked how he and the crew found out the Katu Plantation had been full of Korean laborers, Stephen replied that Americans had planted scouts and spies on New Ireland. It was only after the 65 destroyed the plantation that Americans were able to get word back that there were no Japanese in the plantation at the time of its destruction. Even though Korean laborers were working for the Japanese war effort, the tone of Stephen's voice as he recalled the story made it obvious he felt badly. Korean laborers were enslaved and forced to work against their will by the Japanese during World War II. "They didn't have a choice," Stephen said sadly.

Stephen and the crew of the 65 made it back to Homestead Lagoon aside the LCI 64 on the afternoon of June 7. However, on the trip back, the men noticed something wasn't quite right with their ship. She seemed to be slightly limping, off by just a bit. The next morning, they found out exactly why. After a thorough check of the engines, a General Motors specialist insisted the engines were operating just fine. So a five-man diving crew came aboard to perform a thorough inspection underwater of the starboard screw (propeller) and shaft. About twenty minutes later, a chief named Smith, who was the commander of the divers, reported that one blade had broken off the starboard screw. The ship and crew would just have to manage without a propeller blade until they could get into a dry dock for repairs. The next morning, June 9, the LCI 65 was able to negotiate with another ship to swap some ammo. They traded their armor piercing 40mm ammo in exchange for anti-aircraft 40mm ammo.

———

Later that afternoon a group of four young men, all with the rating of Seaman 2/c, reported aboard for permanent duty aboard the LCI 65. The newest members of the 65's crew were Donald Faleide, Albert Reese, Clyde Reese, and a young man from Hattiesburg, Mississippi, J.R. Reid. After the war, J.R. Reid would outlive every member of the crew and become the last surviving crewmember of the 65.

When reached for an interview in 2014, J.R. Reid spoke at length about his time aboard the LCI 65 and his memories of Stephen and the rest of the crew. But the most comical moment he shared about Stephen wasn't even from the war—it was from a phone conversation that took place decades later, toward the end of Stephen's life, when Stephen rang J.R. up to chat, mistakenly thinking he'd reached the 65's other Reid—James Kent Reid. Although their last names were spelled the same, J.R. was a different Reid, not Stephen's friend, James. Bless his heart, for those who knew Stephen in his elder years, according to loved ones—this sounds exactly like something he would do. J.R. humorously recalled, "James K. Reid, that's who—he was a signalman and they worked close together—and I think that's who [Stephen] thought he was calling. I talked to him for a good while, but he never did recognize that he had the wrong one, I guess." J.R. continued, "I'm pretty sure he was trying to get the signalman from Virginia, because they worked up in the conning tower together. They was on duty at the same time, I think, a lot of times, I imagine. 'Cause Reid was a signalman and your grandpa was a quartermaster so they had to be up in the conning tower when we was underway." James K. Reid had been Stephen's buddy since he became a member of the crew back in March 1944, and now J.R. Reid and three others would join them in making the 65 their permanent home as the war raged on. Since the 65 was no longer a "large" class LCI, the additional space that had been reserved belowdecks for carrying troops was now converted, accommodating the new crewmen.

For the next several days, the crew remained anchored in Homestead Lagoon with routine duties. These would be the 65's last days as officially being classified as a "large" class landing craft. On June 13, the 65 departed Emirau Island with the LCI 64. For the next several days, the two LCIs would sail to Carter City, Florida Island. However, three days into their southern journey, Seaman 2/c Walter W. Henry suddenly began to suffer an acute appendicitis as the 65 passed Simbo Island in the New Georgia Group on June 16. The LCI 64 quickly came alongside the 65 and Henry was carried aboard the 64 on a stretcher, where he would be transferred to Munda Hospital on southern New Georgia. Henry would remain at the U.S. Naval Base Hospital for the next month and a half until the end of July.

————

The day had finally come.

On the morning of June 20, 1944, the LCI 65 was officially renamed to USS LCI (G) 65. She was henceforth officially a gunboat. She would soon have her ramps removed and be reassigned to a gunboat LCI flotilla group for the invasions to come. The 65 was beached at Carter City, Florida Island, with various other landing craft when the ship received orders shortly before 9 a.m. to be received into dry-dock for much needed repairs, including replacing their missing propeller blade. Despite being sleep deprived from their non-stop patrol duties over the past month, Stephen and the rest of the crew were in the mood to celebrate. Their LCI was just rechristened as a brand-new ship after all! The least they could do was crack open a beer from their two-beer allotment and toast to their new gunboat—a proud Navy tradition when a ship got a new name.

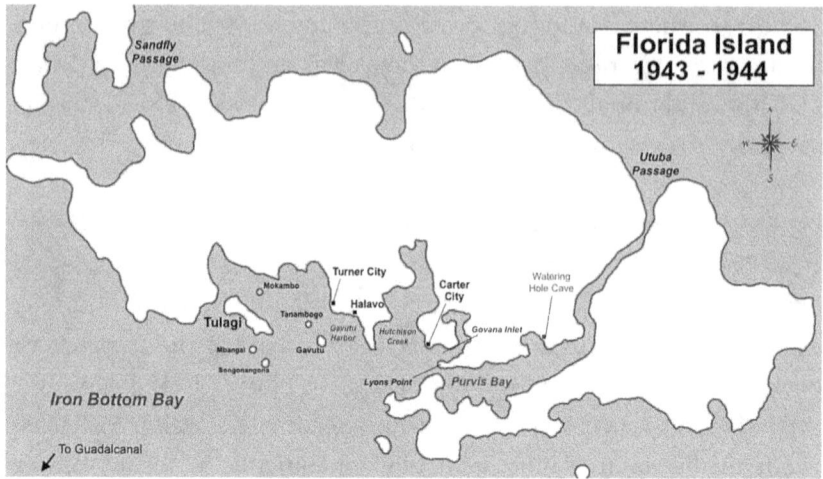

A map of Florida Island, Solomon Islands, June 1944.

But when the men approached the beaches of Carter City, however, the crew of the 65 were surprised and angry when they were denied their beer rations from navy personnel ashore. Stephen remembered after spending that previous month patrolling near Emirau Island that the exhausted men tried desperately to get some beer. "That's for thirty days," he said, "and when we come back, we asked them for our beer allotment, and he wouldn't give it to us." It reminded Stephen of the same situation that happened after the month-long patrol of Rendova in August 1943 aboard the LCI 329, when he was also denied his beer ration. J.R. Reid remembered all too well, "we didn't get it very often." Another LCI veteran, William H. McCracken of the LCI (R) 1030, explained further that "Beer, in the U.S. Navy Pacific theater, was carefully controlled, protected, guarded, regulated, preserved, and rationed by the big ships, and, despite its tremendous role as a morale-builder, in the end the LCIs —at the bottom of the totem pole—got like nothing."[7]

Stephen and the crew were expecting some sort of celebratory gesture to be made from their skipper Macaluso at any moment. Maybe he'd surprise them for such a special occasion. But from the moment they dry-docked, Macaluso had put the crew to work, making them scrape and paint the ship's bottom and sides on a particularly hot, humid day. Once that aching, grungy job was finished, they were immediately put to work installing the starboard screw. The men finished at precisely 5 p.m. Still, they waited eagerly hoping their allotment of two beers per sailor would finally be distributed now that the workday was officially over. But no beer came. Odd. Maybe the next day would be the day.

However, the next day came and went and all it brought was more welding duties. Their main engine exhaust pipes were finally replaced, which was great news, but no beer. But when the next day arrived with no beer and news that they'd be departing from dry dock first thing the next morning, Stephen and the crew hit a breaking point. They realized Macaluso had no intention of christening their new ship with a toast, and worst of all they wouldn't be getting any booze, period. The men were in no mood to be deprived of their rightful beer after the last month and a half they'd had. They had set their expectations too high to be told no. So, Stephen, Oberson, DeWitte, Reeder, and James K. Reid hatched a plan to get their precious beer from the dry dock supply. They were going to steal it.

J.R. Reid remembered four out of the five conspirators, "Well, [Stephen] and Oberson and DeWitte . . . He was in on it . . . Reeder . . . he was in on it. I think there was five of them. And that'd be Reeder, your grandfather, Oberson, DeWitte and..." he trailed off trying to remember the alleged last man, James K. Reid.

The crew received word that they would be departing from dry dock at 8 a.m. sharp on June 23. The men figured as soon as they were out of dry dock fully repaired, they would hastily be off to their next destination in the South Pacific. Stephen and the men gambled that by the time the men working aboard the dry dock realized any beer had been stolen, their gunboat would be long gone. Maybe they'd be sent right back to the Bismarck Archipelago for patrol missions off Emirau Island. Or maybe they'd be sent to New Guinea to help the Allied push there. But surely the men thought they'd be far away from the crime scene. Working together and undetected, five to seven men managed to steal fifteen cases of beer from Landing Craft Repair Unit No. 1 before departing. As part of the plan, Stephen remained aboard the 65 in case any of the officers came looking for the missing men. As Stephen remembered it, "I wasn't the one that loaded it up on the ship, I wasn't the one that put it up in the forward box."

At 7:50 a.m., the LCI 65 made all preparations to depart the dry dock. Twenty-seven minutes later, the ship was waterborne once again with its stolen cargo aboard. The fifteen cases were hidden in the void tank of the boatswain's locker at the front of the ship. However, almost immediately their plan began to fall apart. Though the ship was now fully repaired and sailing out of dry dock, it was not leaving Florida Island. The nervous men were told they'd be beaching at Carter City instead! Macaluso went ashore that evening to get instructions on their next duties. The men grew more nervous when Macaluso returned a few hours later to inform them that they'd remain beached at Florida Island for the next few days.

Stephen and the other thieves sat beached at Carter City as their nightmarish scenario unfolded before their eyes. Instead of sailing far away from Landing Craft Repair Unit No. 1, the LCI 65 was plopped there like a sitting duck on the beach within eyeshot of the dry dock. Each time the men looked at the dry dock it was a reminder of what they had just done, and what they were secretly carrying aboard. With each passing day, the men grew more and more nervous of getting caught. After three excruciating days of anxiously waiting to depart Florida Island, and with no orders to leave, the crew decided to try something desperate. Even though they were idle and under close supervision by the officers, the thieves attempted to get rid of the evidence—by drinking it.

It was around this time that Chief Petty Officer V. F. Herbert was made aware by his storekeepers ashore that fifteen cases of beer had suddenly gone missing from their dry dock. According to inventory, someone had to have swiped the beer within the last few days. As Chief Herbert began to get to the bottom of it, he hoped the perpetrators were still nearby.

And that's when it happened. Sometime near the end of that three-day period, Stephen and Oberson were caught red handed as they were drinking the beer. "I got pulled into it, because I drank a beer," Stephen admitted. It soon became clear to the rest of the crew that Stephen and Oberson would have to take the fall for the theft of the beer because of Paul DeWitte. When asked if the other men pinned the crime on him, Stephen wisely cracked, "Oh yeah," with a smirk on his face. J.R. Reid explained why this was the case, saying, "A chief petty officer on the island over there, I don't know how he got word, about who got it or anything, but he came over and he got them all together—and this DeWitte was in the regular navy and he'd been in trouble three or four times. And [Herbert] got [DeWitte] and he said, 'if you don't tell me who was in on it, what ya'll did with the beer, you're gonna be kicked out with a dishonorable discharge.' So [DeWitte] got together with them and told 'em what was gonna happen to him and they agreed to go ahead and let him talk." Reid laughed as he remembered, "That's the way they got caught!"

Stephen and Oberson's fates were sealed the moment Chief Herbert came aboard the 65 with a group of officers to search the vessel for the missing beer on the morning of June 27. It didn't take them long to find the fifteen cases hidden away in the boatswain's locker. Chief Herbert and his officers reclaimed what was left of their precious missing items and returned to Landing Craft Repair Unit No. 1 on the beach.

Two days later, while still beached at Carter City, the two unlucky boys from Detroit were tried and convicted at deck court, also known as a "captain's mast." Stephen and the gunner's mate Oberson listened to the skipper read the charges of "stealing of property of the U.S. government intended for the Naval Service thereof..." Stephen was found guilty of stealing four of the fifteen cases, which in those days was worth seven dollars total. As punishment the two men would be demoted to the next inferior rating. Stephen would be bumped down from Quartermaster 3/c to a Seaman 1/c. For all the trouble they got into, J.R. Reid humorously quipped, "I guess they just wanted to have it to drink when we left."

For some reason, for the remainder of his life, Stephen wore a tiny badge of shame over the beer-stealing incident. This was despite his future family finding the story hilarious each time he'd retell it over many Thanksgiving and Christmas dinners. Perhaps it was because this incident, as well as the censorship violation he had been hit with in January 1943, would ensure that no matter what happened for the remainder of the war, Stephen would never qualify for the Good Conduct Medal. Or maybe it was the fact that he was never one of the men who actually stole the beer from the dry dock—or hid it —but was found guilty as if he had been. Whatever the reason, it will forever remain a mystery.

————

On the last day of June 1944, the final piece of the legacy large landing craft was removed from the ship. After taking on ammunition from Tulagi, the newly named gunboat moored alongside the docks of Carter City. In less than twenty minutes, cranes slowly amputated the 65's ramps from each side and lifted them off her. Without its distinctive ramps used to disgorge troops onto enemy beaches, the newly named and newly armed LCI was completely transformed. On that hot, muggy day in the South Pacific, the young crew from all over the U.S. watched as the cranes put the finishing touches on their conversion. After quite a journey, their months-long metamorphosis into an LCI gunboat was finally complete.

13

JULY 1944: THE SEVENTH 'PHIBS

"I love the infantry because they are the underdogs. They are the mud-rain-and-wind boys. They have no comforts, and they even learn to live without the necessities. And in the end they are the guys that wars can't be won without."

– Ernie Pyle[1]

The diesel engines of the LCI 65 roared to life on the second day of July, just as the sun was rising. It was finally time to get going out of the Govana Inlet at Florida Island. Macaluso had received orders for their first special patrol mission as an official LCI gunboat. The 65 followed behind the LCI 64 as they both steamed northwest toward their destination of Bougainville.

That same morning in a different corner of the Pacific the Allies had begun the invasion of their next target in the Southwest Pacific: Noemfoor Island, Dutch New Guinea. Since Admiral Halsey's South Pacific Force had taken the Solomon Islands and neutralized Rabaul, General MacArthur's Southwest Pacific Force continued moving up the northern coast of New Guinea. By this point in the war, with Operation Cartwheel officially accomplished, MacArthur was now completely focused on recapturing the Philippines, which had been his goal ever since he was forced to evacuate there in early 1942. And the only way he could earn his vindication was by first capturing and controlling Dutch New Guinea and the Dutch East Indies. By the opening days of July 1944, he had moved the front lines to northwest New Guinea.

As the two LCIs approached the rainy coast of Treasury Island the next morning, they had passed Rendova Island just hours before. It had been almost a year to the day since Stephen's first encounter with the enemy and the bloodshed he witnessed aboard the LCI 329 in Rendova Harbor the previous Independence Day. It was approaching the one-year anniversary since the 65 lost one of her precious firemen, Hurley Christian.

Yet the fighting was far from over. The 64 and 65 moored alongside each other in Blanche Harbor and dropped anchor for the remainder of the day. The next morning, July 4, the crew of the 65 departed for Empress Augusta Bay, Bougainville Island, and arrived later that afternoon, where they remained until the next evening. Shortly after 6 p.m. on July 5, Stephen and his shipmates made preparations to get underway for their special patrol mission south of Empress Augusta Bay, in company with fellow gunboat LCI (G) 22. A little over an hour later, all hands were called to battle stations aboard the 65, as the two LCIs arrived off the shoreline of their target area in the southern waters of Empress August Bay. The 65 trailed just behind the 22. Now under the cover of darkness, they began patrolling a mile and a half offshore of Bougainville. Their patrol began from the southern patrol point off the coast of Motupena Point and extended to the northern patrol point west of the Jaba River, where the river empties into the bay. The LCIs would remain at a distance of a mile and a half offshore as they slowly patrolled one sweep of the shoreline through the night. The LCIs completed their patrol mission when they reached the northern patrol point at the mouth of the Jaba River shortly after 5:30 a.m.

on July 6. The result of the patrol was negative—no enemies sighted.

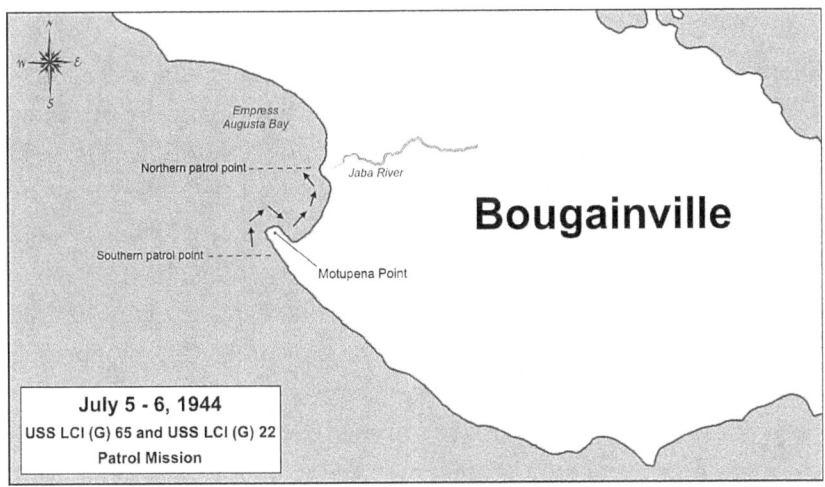

The next morning Macaluso mustered all hands to quarters as the 65 sat anchored in fifteen fathoms of water in Empress Augusta Bay. He had a special announcement for the crew. The men were to receive special instructions for how to request their absentee ballots for the upcoming 1944 presidential election. The current U.S. president, Democratic candidate Franklin Roosevelt, was widely expected to win an unprecedented fourth term against his Republican challenger, Thomas E. Dewey.

The crew remained in Empress Augusta Bay until July 10, when the skipper received orders for their next patrol mission. Shortly before 3 a.m., the 65 hauled in their anchor and departed for New Ireland forming a column behind the LCIs 22, 69, and 64. As the four LCIs headed northwest to New Ireland, the men looked up into the sky shortly after sunrise to see twelve American B-25 Mitchell bombers flying in formation directly overhead. Several hours later, the crew spotted a life raft on their port side. The 65 broke off their formation to investigate the raft, but upon a closer look, the bottom of the raft was missing. If there had been any survivors in the raft, they were no longer there.

The LCI 65 arrived astern of the LCI 64 off the southwestern coast of New Ireland shortly after midnight on July 11. The crew spotted Dunup Plantation just off the coastline, where their patrol began. After three hours of patrolling the coastline, they arrived at Ulaputur Plantation to the northwest, where they reversed course, arriving at Chinese Plantation twenty minutes later. A little over an hour later, the 65 made visual contact with LCIs 69 and 22, and after several more hours, stopped their engines and reversed their course again. But by 4:30 that afternoon the two LCIs left the patrol formation and headed back to home base, leaving the LCIs 65 and 64 by themselves again. That evening around 8:30 p.m., Stephen and the crew arrived off Chinese Plantation yet again, but this time they stopped to observe the plantation because the crew sighted flashes of light coming from it. The ship slowly followed along the contours of the shoreline for several minutes before stopping their engines and firing a star shell from their three-inch 50 caliber gun to illuminate the sky over New Ireland. After firing two more, the 65 and 64 continued their patrol.

Stephen and the crew had been patrolling for nearly forty-eight straight hours, but still they pressed on. By 2 a.m. on July 12, the two LCIs arrived at Huru Point, and by 4 a.m., they were back at Dunup Plantation. After several minutes, both LCIs opened up firing at their target on the beach with their entire arsenal of guns. By 5 a.m., the 65 had fired hundreds of rounds of ammunition, including nineteen star shells. And like that, both LCIs were off again continuing their patrol.

As the clock struck midnight on July 13, Stephen and the crew had now been patrolling for almost seventy-two hours straight with no rest. The men would be forced to endure nearly another twenty-four hours before given orders to return to Bougainville. By the time Stephen arrived in Empress Augusta Bay and began resting at midnight on July 14, they had undergone a nearly ninety-six-hour patrol off New Ireland, including transit time to and from. Stephen and his shipmates were completely exhausted and mentally drained. Luckily, the men were able to rest for the next few days while also resupplying their freshwater stores.

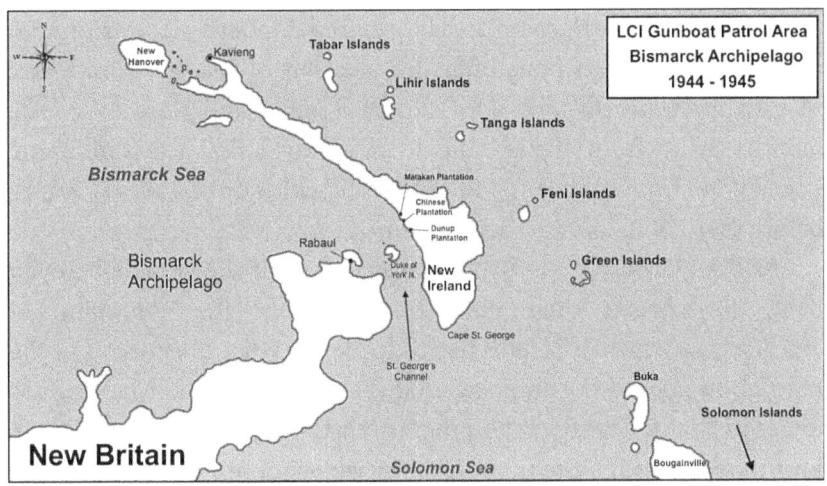

However, on July 17, Stephen's buddies Henry Reeder and Paul DeWitte were once again caught stealing, along with two others—William Harmon and Robert Thornton. All four men were found guilty at deck court and reduced to the next inferior rating. None of the crewmen could recall exactly what was stolen and records did not specify, but because DeWitte this time could not pin it on someone else, he had to take his punishment. But luckily for him, it would not include a dishonorable discharge.

———

July 18 was a special day for the LCI 65. That evening, the men departed Bougainville with the LCIs 23 and 70 for Alexishafen, New Guinea, under a brand-new command. As mentioned, the LCIs of Flotilla Five had been under the Third Amphibious Force, part of the South Pacific Area of operations, under the command of Admiral William "Bull" Halsey. With Operation Cartwheel coming to an official end, Stephen's ship would now be transferred to the Seventh Amphibious Force, under the overall command of Army General Douglas MacArthur.

By July 1944, Admiral Chester Nimitz's Allied forces were in the midst of capturing the Mariana Islands in the Central Pacific theater to MacArthur's north. Nimitz had secured Tinian and Saipan, and was days away from launching the invasion of Guam. Nimitz and MacArthur were pushing ever closer to Japan and were now only months away from being able to begin invading the Philippine Islands. But both men still had a small checklist of islands to capture before they could launch their joint invasion.

As the LCIs 65, 23, and 70 departed Bougainville shortly before 7 p.m., they began their first journey under the command of MacArthur's Seventh Fleet. They didn't know what to expect, but the scuttlebutt aboard the ship was that they'd be added to MacArthur's forces in New Guinea and helping his upcoming push into the Dutch East Indies. Though Stephen would never again serve under Admiral Halsey—and despite a careless mistake committed by Halsey later in October—Stephen remained a dedicated fan of the "Bull" for the remainder of his life.

For the next three days, Stephen and the crew sailed toward Alex-ishafen, New Guinea. Located on the northeastern coast of southeast New Guinea, Alexishafen is just north of a major deep-water port named Madang that had been used by the Japanese as a forward base to protect Rabaul. Between February and April 1944, the Allies invaded and secured both Madang and Alexishafen. Now it was the Allies' turn to use it as a forward base against the Japanese.

On the morning of Friday, July 21, the 65 arrived and anchored in Bostrem Bay with LCI (G) 23 and LCI (L) 230. The crew reset the ship's clocks due to the time zone change and Macaluso went ashore to get further orders. Meanwhile, over twelve hundred miles away at the exact same time in the Central Pacific, Vice Admiral Raymond A. Spruance and Marine Lieutenant General Holland M. "Howlin' Mad" Smith began the bloody invasion of Guam in the Mariana Islands. Like the battles of Saipan and Tinian, the battle to retake Guam from Japan proved to be among the bloodiest battles of World War II, resulting in three thousand killed and more than seven thousand wounded.

The next morning, two Seabees from the 91st Construction Battalion came aboard the LCI 65 for some minor repairs and quickly wrapped up their work by that early Saturday afternoon. The crew spent the rest of the day bringing aboard fresh water and provisions. Stephen was surely looking forward to the next day because it meant that he'd at least be able to go ashore for church services, if only for just a little while. Spending so much of their lives aboard such a tiny ship that felt every wave made the young men relish the moments spent on dry land.

When Sunday morning arrived, Stephen and the rest of those in the church party were permitted to go ashore at 9:50 a.m. As they made their way over to where the church services were to take place, the men were taking their first steps on the tropical soil of the giant island of New Guinea. The young men would only have less than two hours on dry land, but in that short amount of time given, they prayed. And thought. Mostly of loved ones back home. Others, like Stephen, prayed and thought of their brothers doing the exact same thing they were doing, off fighting the war in different parts of the world. The services never seemed to last long enough, and the crew was back aboard the ship by 11:40 a.m.

The next days unfolded much in the same fashion, beached at Bostrem Bay, New Guinea. However, on July 26, two very notable things happened. The first, was a captain's mast that was held for Stephen's troublemaking buddy, Coxswain Henry Reeder, and Gunner's Mate 2/c, Emory Young. Reeder, who had been involved in the beer stealing incident with Stephen a month before, was charged with "insolence and drunkenness." His punishment: "Solitary confinement on bread and water for a period of five days with full rations the noon of the third day." Young was charged with taking part in "Conduct to the prejudice of good order and discipline." His punishment was a demotion to Gunners Mate 3/c, the next inferior rating. When asked how someone could be kept in solitary confinement on such a small ship, J.R. Reid said, "They had a little cage back there, like, if they wanted to put the stores out and didn't want anyone to get in, they'd just take the stores out and put them in it." He went on to describe the small size of the cage as having wiring around it and being "about four feet by six or eight feet, it was very small." He laughed and added that a man in the small cage could "sleep or could stand up [and] walk around, but that was about it!" The second noteworthy event was the return of Seaman 2/c Walter Henry, who had spent the last month and a half being treated at U.S. Naval Base Hospital Eleven for his acute appendicitis that had suddenly occurred on the 65's journey from Emirau to Carter City the month before. Freshly rested and ready to get back into the fight, Henry was glad to see his friends again. Henry, however, had no way of knowing that his reunion with his buddies would be short-lived. It was not the last time he'd have to be treated at a U.S. naval hospital.

USS LCI (G) 69 officers: (left to right) Lt. Commander Archie Holmes (aka "Squad Don"), Arthur Seale, John Ehlers and Lt. (jg) Herman Roesti. Note the artwork—painted by Clarence Johnson—on the front of the three-inch 50 caliber gun tub that they are standing inside. (navsource.org; photo and identification contributed by Desmond H. Johnson)

On the morning of July 28, William E. Harmon, now an Apprentice Seaman as punishment from his deck court on July 17 where he was demoted in rating, was transferred to Medical Facilities Navy 122 for treatment. Ten minutes after Harmon left the ship, Lieutenant

Commander Archie M. Holmes, the commanding officer of Stephen's new LCI (G) Group Forty-Five, now under LCI (L) Flotilla Fifteen, came aboard the 65 to live for a couple days. As the 65 had just transferred to the Seventh 'Phibs, this was the first time Holmes came aboard the 65, but it certainly wasn't going to be the last. As their LCI Group's commanding officer, Holmes would make frequent visits to the 65 in the months to come. Two days later, the Lt. Commander left the 65 and headed back to his LCI Group's flagship, LCI 69. Holmes's men aboard LCI 69 had nicknamed him "Squad Don."

Even funny things can happen during war. On the last day of July, Stephen's buddy Henry Reeder was set to be released from solitary confinement where he had been held in the cage at the back of the ship, as punishment for his "drunkenness and insolence" charge. At 8:25 a.m., Macaluso ordered him to be released and restored to duty from the July 26 incident. However, in a most bizarre turn of events, Reeder didn't even make it a full two hours before he found himself in trouble with the skipper yet again. Shortly after 10 a.m., Macaluso held another captain's mast where Reeder stood alone, front and center, charged with "failure to obey orders." The troublemaking Reeder was thrown right back into solitary confinement for a period of five days. Again, he was sentenced to a diet of bread and water with a full day's ration on the third day. In July alone, this was Reeder's third charge from three different captain's masts, and if you count getting caught in the beer stealing incident from the end of June, it was his fourth offense in the last month. As J.R. Reid remembered it, "He did something later on they put him in the brig for. I don't know what it was for. I guess insubordination." Reid laughed and admitted as he read over the ship's records, "I didn't like the guy anyway, I don't know what happened to him after that." He added, "They've got over here 'solitary confinement' so I don't know what he did."

Macaluso was beginning to lose patience with Reeder.

14

AUGUST 1944: NEW GUINEA

"Caught up in the great necessity of the times, there was a sense of aliveness and purpose that very few things before or since have ever equaled. It was good to have it there to carry us through the drudgery, the boredom punctuated by adrenaline-pumping fright and those just downright bad times. Life was learned much more quickly by our generation as we went through those years."

– John Cummer, LCI 502[1]

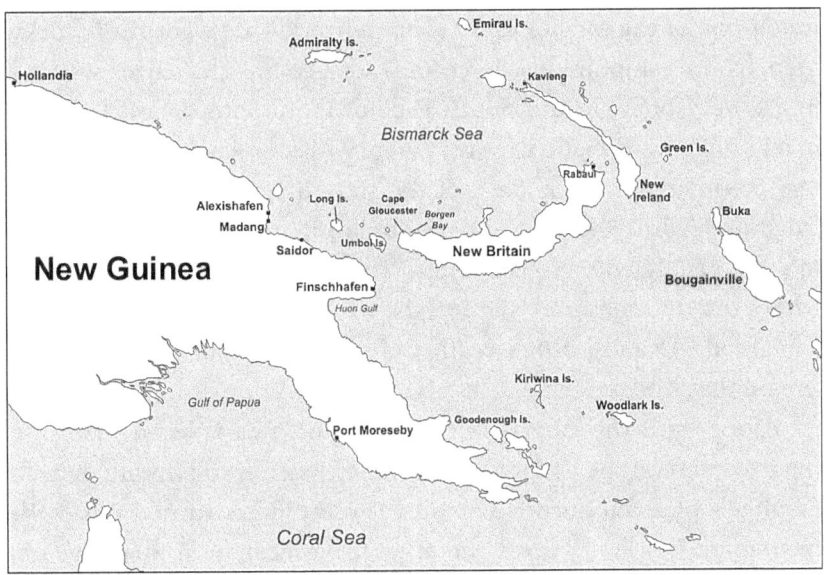

The New Guinea area. Modified version of original map: https://d-maps.com/carte.php?num_car=3336&lang=en

One of the many jobs LCIs had during World War II was to pick up and deliver mail to other ships and locations. This was in its own way a coveted job, as every single American man serving in the war eagerly awaited mail from loved ones back home. Author John Wukovits wrote that mail from home "forged an unbreakable link with loved ones and all that was familiar," and that mail was "a lifeline tossed out from mothers and wives to sons and husbands heading into an uncertain world."[2] But from time to time, an LCI might find itself tasked with other random, non-routine deliveries of equal importance to moral. On August 2, while anchored in Bostrem Bay, the LCI 65 found itself tasked with both types of jobs after a motor whale boat pulled alongside her. The 65 was given orders to deliver mail and additional gear to the supply officer at Navy 722 at Cape Cretin, Finschhafen, New Guinea.

But the 65 was also given something else to deliver: three movie reels. This meant that the crew of the 65 might potentially be able to watch one of the movies upon their delivery if they got really lucky. Men of the Amphibious Force did not get the chance to watch a movie very often in the South Pacific. Word spread fast. Unfortunately, once word made its way to Henry Reeder who was still in solitary confinement at the back of the ship, it was too late. Macaluso, frustrated with Reeder's obscene amount of offenses in the last month, had Reeder transferred to the USS *Rigel* for safe keeping while the 65 delivered the goods to Finschhafen. As J.R. Reid explained, "Usually, if they really did something bad, they'd transfer 'em off the ship somewhere, let him serve the time."

Along with the cargo, four officers and two enlisted men also reported aboard the 65 for the temporary journey to Finschhafen. To Stephen's pleasant surprise, one of those officers turned out to be Lieutenant (jg) Frank Love, the executive officer from Stephen's old ship. Stephen had not seen him since Love transferred to the LCI 64 on New Year's Eve 1943, but he was glad to see Love again, even if it was only for a short time. The convoy arrived in Langemak Bay, Finschhafen by the next evening, and the passengers that the 65 had been carrying, including Love, departed the ship at around 5:45 p.m. Once again, Stephen said his goodbyes to Love, wondering if he would ever see him again. Macaluso also went ashore at Finschhafen on official business and returned ten minutes later.

A devilishly handsome Ensign William E. Keeler just also happened to be in New Guinea around this time. He reported aboard the LCI 448 in August 1944 as a "90-day wonder" after receiving his commission from Northwestern University back on May 10. In his personal memoirs, Keeler recalled a story at Hollandia, New Guinea, in late summer 1944 when his men actually did get to watch a movie ashore:

> Someone set up a movie theater ashore in a wooded area with logs for seats and a screen tied between two trees. The first night we went to the movies it was a real nice change and everyone really enjoyed it. They turned the lights on at the end of the movie, and, to everyone's shock and surprise, there were two Japanese soldiers sitting on

a log in front of the screen. Someone took them in custody and turned them over to the army troops there.[3]

Just the sight of lights in the harbor and movies on the open deck seemed like a return to civilization. In his history, Daniel Barbey fondly recalled, "The films were usually old and often in bad condition, but they were always well attended. Sometimes the same movie would be shown twice in the same evening, but no matter, the audience would sit happily through both shows. It was cooler and pleasanter to sit on topside and watch a stale movie than go below to stifling bunks."[4]

There at Finschhafen, the 65 delivered the movie reels, but unfortunately for Stephen and his shipmates, they were devastated to learn that there would be no movie for them. They'd have to set sail soon instead. The men swapped the mail they'd been carrying for more mail, and that evening, the disappointed men aboard the 65 joined the convoy of LCIs 224, 229, and 29 as they all departed back north to Alexishafen, where they arrived on August 5. That same morning Reeder returned aboard the 65 from the *Rigel* where he was serving out the rest of his five-day sentence.

On the morning of Sunday, August 6, the crew of the 65 did however get to go ashore to attend church services shortly after 8 a.m. The men gathered ashore and again prayed for peace and an end to the war before returning back to the ship forty minutes later. The crew would remain beached in Bostrem Bay, Alexishafen, for the next week, where they took on water and compensated the compasses. On August 10, Seaman 1/c Joseph Orbich was taken to the sick bay aboard the LST 353 for hospitalization and treatment for an unknown injury. The next day, August 11, Lieutenant Commander Archie Holmes again came aboard the 65 along with Lieutenant R. Jordan and two enlisted men to do a routine check that only took five minutes. Later that same day, two radio technicians from the LST 455 came aboard to run tests on the 65's ABK radio transponder, which was reported in good working order.

On August 13, the 65 received orders from the landing craft

control officer at Finschhafen to depart for Cape Gloucester, New Britain, to pick up a passenger who needed transportation from there to Alexishafen. On the way the 65 joined up in a convoy with LCTs 382 and 175. The men exercised fire drills and conducted other routine tasks while en route to New Britain. The 65 arrived at Borgen Bay, New Britain, the next afternoon, beaching at Gillgall Beach, located on the far west end of Borgen Bay, shortly after 4 p.m. The two LCTs also beached nearby, and Macaluso left the ship to go ashore and returned with orders an hour later.

On the morning of August 15, the Australian Port Director from Cape Gloucester came aboard the 65 for transportation to Alexishafen. After having some initial trouble retracting from the beach, two LCMs came over to the 65 to lend assistance. After about twenty minutes, the two LCMs shoved off and the 65 was underway by that afternoon in company with LCTs 172 and 385. The 65 spent the rest of the evening on course for Alexishafen when all of a sudden they received an urgent dispatch in the middle of the night. According to the dispatch, the 65 and the rest of the convoy was to proceed to Point Reumer, Long Island—a tiny, doughnut-shaped island located to the west of New Britain and Umboi Islands—to pick up additional passengers and help a distressed LCT. By sunrise on the morning of August 16, the convoy reached the coastline off Cape Reumur, Long Island. At around 7:45 a.m., the LCIs 65, 69, 64, and 23 beached at Kiau Point, located at the northern coast of the island, in company with LCTs 175 and 382. That afternoon, Macaluso left the ship for an hour to get further instructions on the sudden dispatch they had received. He was told there would be a slight change of plans. That night, the 65 was to load twenty-six additional U.S. Army troops and was also to make a pit stop at Saidor, New Guinea, where the crew would drop the solders off with the Australian Port Director. The Australian they picked up in New Britain would no longer be going to Alexishafen. Right on time at 9 p.m. sharp, the 65 took aboard the twenty-six men under the cover of darkness. The next evening on August 17, Lt. Commander Archie Holmes reported aboard the 65 for temporary duty, where he would assist Macaluso with the drop-off. Later that night, the 65 cast off the beach at Kiau Point and headed southwest for Saidor, with LCI 23 and LCTs 385 and 172.

In the early morning hours of August 18, the 65 entered Saidor Harbor and beached alongside LCT 385 and LCI 23 nearby the Port Director's hut. Shortly after 7:30 a.m., the twenty-six U.S. Army troops, as well as the Australian Port Director from New Britain departed the ship. A half hour later, the 65 secured a pump and transferred about seven hundred gallons of water over to LCT 385. That night, with the convoy's mission completed, Stephen and the crew of the 65 were given clearance to return to Alexishafen. The 65 retracted from the beach and departed Saidor in a column with LCI 23 and LCTs 172 and 385. The 65 arrived back at Alexishafen shortly after 8 a.m. on August 19. After they beached at Bostrem Bay, Lt. Commander Archie Holmes left the ship. At the same time Holmes departed, Joseph Orbich returned aboard after going on sick leave nine days earlier. However, oddly enough, Orbich was forced to leave once again several hours later for the LST 453's sick bay.

That night, three shipfitters from the LST 455 came aboard the 65 and commenced work on various parts of the ship: Shipfitter 1/c Calvin P. Carlyle, Shipfitter 2/c Louis M. Brewer, and a young man from Knoxville, Tennessee, Shipfitter 3/c John Joseph Angle. When they finished their welding work shortly before midnight, the three shipfitters returned to the LST 455. Little did the 65's crew know, this would not be their last encounter with the LST 455. They would meet again in a few months in Leyte Gulf, Philippines, on a fateful day in November, a day that LCI sailors would later nickname "Bloody Sunday." And although John Angle had no way of knowing it, these would be the last few months of his young life.

For the next three days, Stephen's ship remained beached at Bostrem Bay. On the morning of August 22, the LCI 64 came alongside the 65 and dropped off Ensign B. H. Learish, who needed transportation to Hollandia, New Guinea. However, right as the 65 was departing Alexishafen for her two-day journey to Hollandia, while retracting from the beach, LST 18, which was located right behind the 65 on her port quarter, asked the 65 if she was coming off the beach, in which the 65 replied in the affirmative and asked LST 18 to stand clear. Unfortunately, due to a mechanical malfunction the LST failed to stop, and as a result, the stern of the 65 struck it on the port side, which ripped the LST's seams to the extent of about two feet. The 65 suffered only slight damage to an aft stanchion (a sturdy fixture for support) on her starboard side. With that little mishap behind them, the small gunboat limped into the rear-most position of a column formation behind LCI 69 at the lead, followed by LCIs 64 and 23, as they sailed northwest to Hollandia at a speed of nine knots.

The quartet of LCI gunboats arrived at Humbolt Bay in Hollandia, on the morning of August 24. Hollandia, known today as Jayapura, is located on the northern coast of New Guinea. Huge numbers of Allied ships had been gathering off Hollandia for some time in preparation for their invasion of the Dutch East Indies in the coming weeks. Scheduled for mid-September, the LCI 65 and the rest of the Seventh 'Phibs were to take part in the Morotai Island campaign. General MacArthur needed to use Morotai—the last steppingstone of his Southwest Pacific island-hopping campaign—as a forward base for his massive, long-anticipated invasion of the Philippine Islands, which was scheduled for late October. The Allies would use the last weeks of August and first weeks of September to rehearse and prepare for Morotai. But first, Stephen's banged-up LCI needed some repairs.

The 65 moored alongside the LCI 64 about thirty miles off the coast of Hollandia, and Ensign B. H. Learish departed from the ship, which concluded his two-day trip with the 65. The next day, August 25, was Stephen's twentieth birthday. On that day, the commanding officer of the LCI Flotilla, Lt. Commander Archie Holmes came aboard the 65 for a ten-minute inspection in the early afternoon. Lieutenant Kateci, a civilian radar technician, also came aboard to work on the 65's radar for the rest of the day, before departing from the ship at 7 p.m. sharp.

The next day, August 26, the 65 received orders from the commander of LCI (G) Group Forty-Five to place a grapnel hook over the stern of the ship and commence sweeping for enemy submarines which were reported in the area. For the next three hours, the 65 swept the waters off Hollandia, and after a thorough search, the 65 reported finding no Japanese submarines. After departing the sweep zone and dropping anchor that evening, Joseph Orbich reported back aboard from the sick bay of the LST 453.

On the morning of August 27, Macaluso and his quartermaster, Stephen, left the ship and headed for the USS *Blue Ridge* to secure charts for the upcoming invasion. The *Blue Ridge*, an Appalachian-class amphibious force flagship, had been made the flagship of Rear Admiral Daniel E. "Uncle Dan" Barbey in December 1943. Barbey, a brilliant military tactician and planner, commanded the Seventh 'Phibs aboard the *Blue Ridge*. One can only imagine Stephen's reaction upon seeing Uncle Dan in the flesh. Aboard his flagship, Barbey had been preparing and planning for the amphibious invasion of Morotai.

After securing charts of the area and receiving instructions for the coming days from the Seventh 'Phibs staff, Macaluso and Stephen left the *Blue Ridge* and returned aboard the 65. Stephen and his shipmates would spend the remaining days of August practicing gunnery drills off the northern coast of Hollandia and receiving repairs in preparation for the coming invasion.

15

SEPTEMBER 1944: MOROTAI

"I believe the worst of our war is still to come, and that before it is over everybody in America will really feel it. I hope so, because then the boys overseas won't feel so lonesome."

– Ernie Pyle[1]

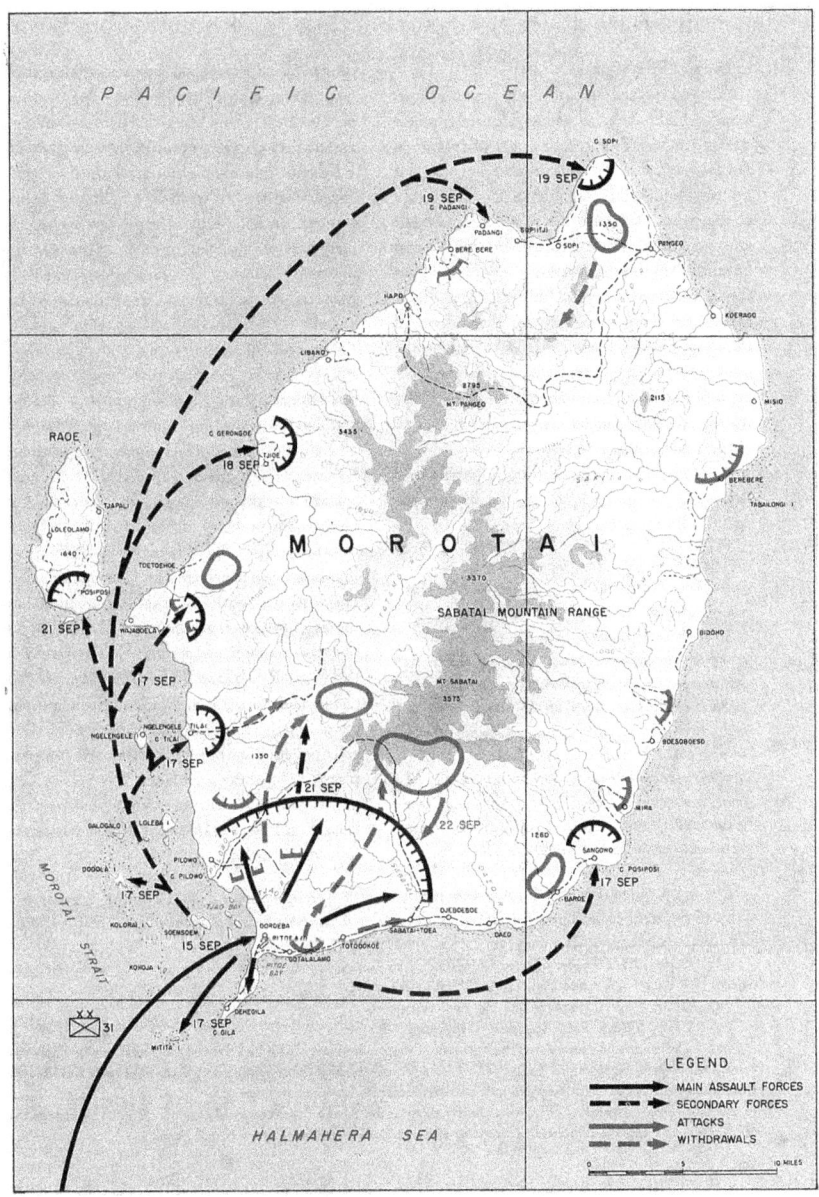

A map of the movements of Allied and Japanese forces during the first weeks of the Battle of Morotai. From *The Campaigns of MacArthur in the Pacific* (1966), reports from General Douglas MacArthur, p. 175.

"You see this, Halmahera, you see that little bitty island right there? That's what they call Morotai Island. I invaded that one too."

At eighty-six years old, Stephen's hands reflected the lifetime of work he'd spent raising a family as a handyman and sheet metal worker. They looked harsh and worn. His nails looked rough and craggy from years of hard labor and carpentry. Hands of someone who lived through the Great Depression and did mechanical work. The thumb on his right hand barely moved as it slowly pointed to a small group of islands on a map of the southwest Pacific.

"It's right down here, this little bitty guy here." Stephen paused. "That was Morotai."

Morotai—a small island forty-four miles long and twenty-five miles wide located about twelve miles across the channel from the northernmost part of the much larger nearby island of Halmahera—was part of the Molucca Islands group during World War II. Today, the archipelago is known as the Maluku Islands, and is part of eastern Indonesia. However, in 1944 Morotai was just another island the Allies needed to secure in order to eliminate Japanese threats to their coming invasion of the Philippines. The Moluccas lie right smack in the middle between the bird's-head–shaped Vogelkop part of northwest New Guinea and the southernmost Philippine Island of Mindanao. Geographically, this was a dangerous spot for the Japanese to control, as Morotai's enemy-held airfields were a threat to all sea traffic and Allied shipping to the Philippines.

Morotai was chosen for invasion due to intelligence reports that claimed there was nothing more on the island than five hundred Japanese troops and an airfield that was started but never finished. Halmahera, on the other hand, according to U.S. Naval historian Samuel Eliot Morison, was estimated to have about thirty-seven thousand Japanese troops, as well as eight or nine airfields. Such a strongly held island, Morison argued, "would not be worth the cost."[2] Thus, Morotai was chosen instead of Halmahera as the better target. And there was only one suitable spot for a landing. Since Morotai is mountainous and covered in forest, Gila Peninsula—the tiny level peninsula at the southern tip of the island—was selected for the invasion, as it presented the most favorable conditions for quick airfield development.

Stephen motioned again on the map of New Guinea, sliding his right thumb from Alexishafen to Hollandia as he continued, "Well we went through the northern part of New Guinea . . . we stopped off here and rested up, and got reshaped, you know?" He moved his thumb in a northwesterly direction, up to the Molucca Island group as he spoke, "but then we come out of there and went right to Morotai Island just to give us an anchor and an airfield."

In the early morning hours of September 6, 1944, a massive preliminary rehearsal for the Morotai landing began to take place for the Seventh 'Phibs off the northern coast of New Guinea at Maffin Bay. At around 4:30 a.m. local time, Stephen and the crew of the 65 fired up the main engines and got underway to join the massive armada of various other ships and landing craft forming up in the darkness. The 65 fell into a column in which they followed behind the LCI (G)s 69, 23, and 64, and ahead of the rocket equipped LCI (R)s 31 and 34. Over the next five days, the ships conducted gunnery drills and practiced different coastal bombardment maneuvers. The 65 also took on thousands of gallons of water before the invasion.

Then, on the morning of September 11, the day had arrived. Pursuant to orders from Operational Task Unit 77.3.2. from the Commander of the Seventh Fleet, Macaluso and the crew made all preparations to get the gunboat underway for invasion to execute Operational Plan 8-44. The LCI (G) 65 was part of CTU 77.3.28, which included the following vessels during the landing on Morotai:

Starboard side of 65:
USS LCI (G) 69 (Group Flagship)
USS LCI (G) 23

Port Side of 65:
USS LCI (R) 31
USS LCI (R) 34

Ahead of 65:
USS LCI (G) 64
USS LCI (G) 70
USS LCI (G) 68

The Allied convoy of American and Australian ships that dotted the horizon steamed toward the Dutch East Indies at about five knots. It was a journey that would take them four days, in which they'd arrive just in time for D-Day on the morning of September 15. To eliminate any potential air threats from Halmahera and other nearby Japanese-held islands in all directions, the U.S. Fifth Air Force and aircraft carriers from the Pacific Fleet conducted bombing raids that were deadly effective. By the morning of invasion, the Americans had managed to knock out almost every workable plane in any place in which the Japanese could strike back. According to Barbey, it was later learned that the Japanese had withdrawn most of their workable planes to Luzon, Philippines.[3]

As the quartermaster, Stephen's battle station was in the pilothouse within the square conning tower. His gunboat took a position on the right flank of the convoy astern (behind) the assault LSTs. The USS *Carter Hall* was also up ahead of the 65's position. The convoy experienced a light shower later that day, but other than that, the weather was clear and the sea calm for the remainder of the journey.

———

A two-fold attack plan was to take place on September 15, 1944. The same day that MacArthur's Seventh 'Phibs and Stephen's LCI (G) 65 were scheduled to invade Morotai, Admiral Chester Nimitz's Central Pacific forces would simultaneously invade Peleliu Island, located in the Palau group of islands, four hundred seventy miles northeast of Morotai. Nimitz's forces had spent the summer capturing Guam and the Northern Mariana Islands, and it was now time for the 1st Marine Division to slug it out with the Japanese on Peleliu. The Old Breed of the 1st Division would fight one of the deadliest, bloodiest, most brutal battles in World War II. Military brass predicted it would only take the Marines three days to capture, but Peleliu ended up taking over two months. Prior to the battle, American intelligence had a wildly inaccurate assessment of Japanese forces defending the island in every way. Aerial photographs completely missed the hidden Japanese dug into the Umurbrogol mountains of Peleliu. As the Americans inched closer and closer to their homeland, the Japanese had spent over a year digging into the limestone caves within the cliffs and hills preparing for a brand-new defensive strategy—the same one they would eventually use at Iwo Jima and Okinawa. No more banzai charge attacks that were wasteful and ineffective. It was now going to be a war of attrition. Under the command of Colonel Kunio Nakagawa, the Japanese would fight nearly to the last man from within the interconnecting and interlocking caves that were meant to inflict maximum casualties. The Marines were completely unaware of the coming slaughter. The initial amphibious landings were horrific, as Commodore John H. Morrill witnessed firsthand aboard LCI 730. The extreme heat was well over 100 degrees Fahrenheit and supply problems prevented the Marines from getting fresh drinking water for days.

In terms of deaths per number of fighting men, the Battle of Peleliu resulted in the highest casualty rate of the Pacific War.[4] The decision to invade Peleliu has become a controversial one in the eyes of history, due to its questionable strategic value. Peleliu's capture was supposed to aid and cover MacArthur's eastern (right) flank in his invasion of the Philippines, however, historians would overwhelmingly conclude that the seizure of Peleliu proved ultimately pointless in that endeavor. One need look no further than the fact that a month later, on October 20, when MacArthur's forces landed on Leyte, the 1st Marine Division was still in the thick of savage fighting on Peleliu with no end in sight.

———

Stern (rear) view of USS LCI (G) 65 and USS LCI (G) 64 (left), underway at
Morotai in September 1944 during the Western New Guinea operation. Based
on deck log records, the two LCI (G)s in front of LCI (G) 65 may be USS LCI (G)
23 and USS LCI (G) 70. (navsource.org)

It started right on time. Pre-dawn. Suddenly, from within the conning
tower, Stephen could hear the bombardment of Morotai begin in the
early morning twilight. As the fierce little gunboat approached Cape
Gila on the southern tip of the island, the heavy guns of the
destroyers and Australian heavy cruisers grew louder. Starting at
around 6 a.m. on the morning of September 15, the deafening
bombardment and shelling from the massive Allied ships began
raining hellfire down on the island for two straight hours. With pits
in their stomachs, the crew watched as tracer shells ripped and
streaked through the pale blue darkness as each one pulverized the
island.

Once the landings started the gunboats fanned out into their assigned areas. In pairs, the LCIs were ordered to cover the areas around Red Beach to the north and White Beach to the south. The beaches were located on each side of the Pitoe airstrip. The LCI 65 and LCI 64 would together patrol the bombardment zone between Red and White Beaches and provide covering fire as the amtracs (LVTs) landed troops on the beach. LCIs 69 and 23 were to cover the left flank of Red Beach. LCIs 70 and 68 were to cover the right flank of White Beach.

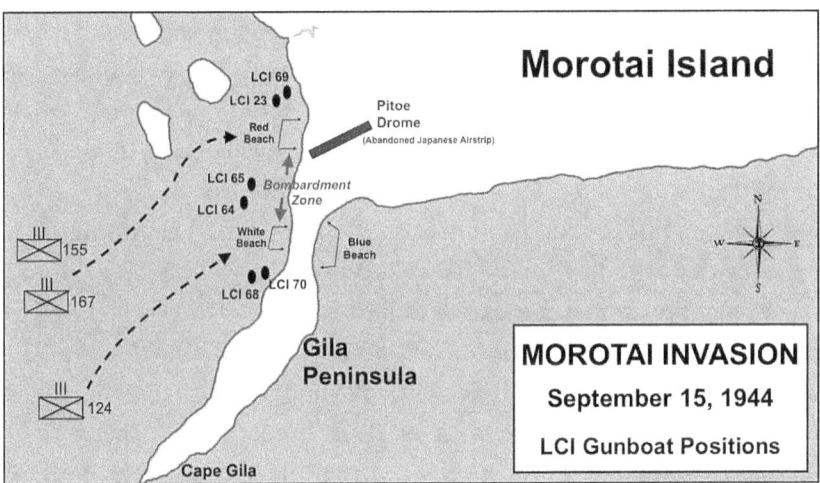

The invasion forces were made up of about 28,000 combat troops from the XI Corps under the command of Major General Charles P. Hall. The heart of this force was made up of the Army's 31st Infantry Division under the command of Major General John Persons. Admiral Barbey was now commanding all naval operations of the Seventh 'Phibs aboard a different ship, USS *Wasatch,* with General MacArthur observing from the cruiser *Nashville.* The U.S. Fifth Air Force, as previously mentioned, provided the air cover during the landings, and had complete supremacy over the skies. Once the initial bombardment ceased, the 155th and 167th Regimental Combat Teams (RCT) would land ashore at Red Beach to the north of Pitoe airstrip, and the 124th Regimental Combat Team would land at White Beach to the south of the airstrip. Both regiments' D-Day objectives would be to capture the airfield, then move inland to establish a perimeter around it.

Shortly before 8 a.m., the LCI 64, alongside Stephen's LCI 65, arrived in their position off Red Beach. However, since this was the 65's first official invasion as an LCI gunboat and since she no longer had her ramps to deliver troops, her orders were to utilize her new armament of machine guns. As quickly as it began, the heavy bombardment from Allied destroyers and cruisers suddenly ceased. But the Allies had learned from their disastrous mistakes at Tarawa in November 1943. There would no longer be a lull period for the Japanese to regroup and reorganize once the heavy bombardment stopped. This was the exact moment the 65 had spent its early months of 1944 preparing for. This was the moment of truth for the rocket armed LCIs and LCI gunboats. As soon as the heavy bombardment ceased, the LCI gunboats unleashed a fury of covering fire as the LVTs approached the line of departure for H-Hour scheduled to begin at exactly 8:30. The 65 opened up with her three-inch 50 caliber and 40mm machine guns. Her crew blasted away at the vicinity of Red Beach, concentrating most of their fire on Red Beach's right flank. A minute later, they reversed course and began firing with all their guns. The crew hit Red Beach with everything they had, and the Japanese offered no resistance. According to Morison's history, "Then came two waves of LCI," referring to the other LCIs that still delivered troops. "The troops carried by these 'Elsie Items' had to wade ashore in water armpit-deep over a very uneven bottom", he wrote.[5] However, unlike the Marines at Peleliu, the Allied landings on D-Day at Morotai would eventually go unopposed—a most pleasant surprise.

The 65 reversed its course again and headed toward White Beach.

Then, a few minutes after 8 a.m., things got a bit chaotic. The heavy bombardment by the Australian heavy cruisers and American destroyers suddenly began again, but they were only targeting White Beach. The 65 found itself in a position between Red and White

Beaches only five hundred yards from shore at the same time the heavy shelling was hitting White Beach. The 65 had orders to be bounded to the left flank of White Beach, so they were extremely close to the heavy bombardment coming from their own ships further offshore. The 65's crew could hear shells whizzing through the air and exploding as they hit the island. Trees and huts burst into flames oozing pummels of smoke. Fragments from the explosions began landing near the 65, but the LCI had orders to remain at her assigned spot near the bombardment zone. One large piece of debris barely missed the ship by only about three feet. Luckily, despite close calls the 65 would not get hit.

USS LCI (G) 64 (front) and USS LCI (G) 65 (near background) attack the area between White and Red Beaches at Morotai, on September 15, 1944, during the Western New Guinea operation. (U.S. National Archives photo No. 80-G-181441)

The rocket equipped LCI (R)s of the invasion were getting ready for their own bombardment of the beach. With their ramps removed also, these rocket LCIs had installed rocket launchers on each side that were armed with 4.5-inch rockets loaded into racks. Their job was to provide covering fire for the landings with destructive rockets to launch at the enemy like bazookas. Just as these rocket LCIs were moving into position, an explosion occurred on the number four gun at the back of the 65. According to Macaluso's official action report detailing the events, "Our numbers 1 and 4 20MM guns each had a (jam) on D-Day and before those jams could be cleared, both 'cooked off' in the guns." The skipper went on, "One of these 'cook offs' wounded a loader."[6]

Seaman 1/c Walter W. Henry in World War II. Henry was awarded the Purple Heart for his injuries sustained at Morotai Island on September 15, 1944. (*Elsie Item*, April 2016, p. 19; Courtesy Ken Howdeshell)

At the rear of the ship, Seaman 1/c Walter Henry was loading rounds into the number-four 20mm machine gun on the starboard side when the gun unexpectedly exploded. According to records, Henry was attempting to clear a jam and suffered an injury from two pieces of shrapnel that struck his right wrist. Years later, J.R. Reid vividly recalled the incident with Henry saying, "He was a loader on a 20-millimeter and when he put the magazine up there it exploded and messed his arm up." Reid went on to explain: "There's a certain way that you—if you didn't get the thing just right, one round would probably explode on you." To make matters worse, Henry had placed his hand in the worst possible position right before the explosion. As Macaluso put it, Henry had vulnerably "placed his hand in position to catch the projectile as it was cleared."[7] J.R. Reid added, "and I think that's all that exploded was one round . . . But it made a mess of his arm I know that. He didn't lose it, but I don't think he could use it after they got it patched up." Henry was beginning to go into shock due to the loss of blood.

Walter Henry's grandson, Ken Howdeshell, confirmed the details of Reid's story in an e-mail from December 2012, stating, "The gun my grandfather was firing during the landing (20 MM I think....) exploded filling his floatation vest, stomach, right arm and hand with shrapnel." Due to Henry's injury, his grandson wrote that "he was slightly disabled in his right hand. Couldn't fully open his fingers...(as if he was gripping a hand saw and couldn't let go)."[8] Howdeshell later explained how the gun crew "would need to change the barrels after firing so many rounds, [because] the barrels would heat up." He described how "the fighting would be intense and there wouldn't be time to change the barrels, leading to them getting too hot and cooking off rounds etc. Especially when under attack by zeros." He added, "My grandfather recalled dealing with incoming zeros at times during intense firing."[9]

H-Hour was approaching and the rocket LCIs and assault waves of LVTs had begun crossing the line of departure as they approached Red Beach. The skipper withdrew the 65 from the area in order to evacuate Henry, who was losing lots of blood and needed urgent medical attention. Within an hour, Henry was transferred with a pharmacist's mate (medic) to the USS LST 456 to seek further treatment. Walter Winfred Henry's war was officially over. As J.R. Reid put it, "We never did see him anymore. They sent him to a hospital and he never did come back onboard."

Once the pharmacist's mate returned aboard the 65 after transferring Henry, the ship continued its patrols off Morotai. As Barbey explained in his history, just because the D-Day landings went unopposed didn't mean the Americans were in the clear quite yet. They still had to prevent Japanese from reinforcing Morotai from nearby Halmahera Island. At midday, the 65 returned to the beach area to drop anchor and await their nighttime patrol with the other gunboats. By this point in the war the American forces were well aware of the Japanese tactics, which almost always called for reinforcing their troops at night under the cover of darkness—usually by barge. It would be up to the LCI gunboats, including the 65, to patrol the waters between the islands and destroy any Japanese barges or enemy vessels that might attempt to send more troops to Morotai. Several hours later, a correspondent named Blaine Fielder from the *Australian Evening Press* came aboard for observation of the night patrol.

Soon thereafter, the 65 joined up with the group's flagship, LCI 69, and the LCI 64 between Roe and Halmahera Islands and got underway. Like the 65, Lt. Commander Archie Holmes had encountered his own set of misfortunes earlier that day aboard his flagship.

"Morotai I remember quite well," LCI 69 veteran Desmond H. Johnson declared with laughter, "because we hit a reef." He continued, "and we were stuck on the reef for most of the invasion. And we had to be pulled off by a tug that finally came in and got us off." Johnson, then a Seaman 2/c, added, "The thing I remember about it so much is that it's amazing how much of the noise of the guns is absorbed by the water—when the ship's sitting in the water. But when it's sitting on a reef, on dry land, the noise is just unbearable."

As the sun dipped below the ocean, the LCI gunboats began their patrols off the southwest coast of Morotai, which would last all through the night and into the next morning. It would be yet another sleepless night for the young men.

At 5:30 a.m. the next morning the 65 joined up with LCIs 69 and 64 and headed for anchorage. About an hour later, the crew spotted Japanese Zero fighter planes overhead. The men opened fire as the planes flew past them. It remains unknown if any of the Zeros were hit. The men remained at battle stations as the ship anchored in its new spot off the northwest coast of Morotai. Fielder left the ship shortly before noon in a PT boat that came to pick him up. However, the crew would see Fielder again briefly in the days to come. Just after sunset, the 65 joined up again with the LCIs 69 and 64 for another nightly patrol between Morotai, Rao, and Halmahera Islands. The three LCIs headed southwest.

Shortly after midnight on September 17, the 65 made all preparations to get alongside the USS *Oyster Bay*, a navy torpedo boat tender, due to a sick crewmate. Shortly after 1 a.m. a doctor from the *Oyster Bay* came aboard the 65 to check on Seaman 1/c Vern Troy Durbin, who had reported aboard the 65 six months earlier on March 12, the same day as Stephen's buddies Reid and Reeder. It only took the doctor several minutes to determine Durbin needed to be transferred to the *Oyster Bay* for further medical treatment. Durbin and the doctor departed the ship around 1:15 a.m. and the 65 immediately got underway. The men would not see Durbin again until November.

Only ten minutes later, as the 65 steamed ahead in complete darkness with one less crewman, the LCI suddenly crunched to a violent halt. It had run aground on a coral reef. Orders were immediately given to the motor macs in the engine room below to stop all her engines and back up. But she wouldn't budge. Orders were then given to transfer the fuel from the forward to aft of the ship, which took over an hour. This did the trick as the LCI slowly began to move backwards. However, the crew soon noticed the electric steering control was out and no longer working. They would have to manually hand steer the ship until they could get it fixed. The best the LCI could do was to anchor until sunrise off the southwest coast of Cape Gila, Morotai. There, the uneasy, tired, sleep-deprived crew bobbled helplessly in the darkness. Floating. Waiting. The weary men snapped back to attention an hour later as they were called to battle stations in the morning twilight. Within minutes the crew began firing at unidentified enemy planes. Then, as suddenly as the firing began, it stopped again. But the 65 didn't waste any time. With the sun beginning to rise, the LCI immediately got underway to get help.

Luckily, it didn't take long to find a nearby ocean tug, the USS *Quapaw*. Shortly after 7 a.m., the 65 pulled alongside the Abnaki-class ocean tug and requested assistance with repairs to their electric steering gear that had been knocked out. Around lunchtime, Blaine Fielder came aboard once again on official business. A half hour later, he was gone for good. After much exhaustive research at the time of this book's publication, it still remains unknown if Fielder ever published an article in *Australian Evening Press* or anywhere else about his experiences at Morotai aboard the LCI 65.

That afternoon, divers from the *Quapaw* commenced diving operations so they could check the hull of the 65 and her screw. After about twenty minutes of repairs, the divers reported that both the hull and screw were satisfactory and in working condition. By the early evening, the 65 was once again underway to continue its patrol duties, finding the LCI 64 and following behind her as they began another all-night patrol off the northwest coast of Halmahera Island.

In the early morning hours of September 18, the crew of the 65 noticed that heavy guns on the beach began to open fire on enemy planes. Five minutes after they ceased firing, the 65 opened up with her own three-inch 50 caliber heavy gun at the front of the ship. She fired furiously at the two Japanese planes in the distance, but it was quickly becoming clear that the 65 was running critically low on ammunition for the three-inch gun. About an hour later, the crew got underway to get resupplied by another ship. At around 7:30 a.m., the 65 beached on the southwest coast of Morotai and moored alongside the LCI 68, where she commenced taking onboard one hundred forty-six rounds of three-inch 50 caliber ammo. The 65 was off again, until a brief rendezvous with the LCI 23, which also needed some three-inch 50 caliber ammo. The generous crew of the 65 transferred sixty shells over to the LCI 23, and then later got underway to continue patrolling area "Able" with the rocket ship LCI 72.

For the next three days, the 65 conducted routine patrols off Morotai, but in Macaluso's words, "activity was null." He noted in his action report that "the only other action experienced by this vessel came on 21 September 1944."[10]

The night before, on September 20, the LCI 65 had departed from Cape Gila in a convoy of ten other ships, including the LCIs 70 and 73.

The LCI 65 had orders to deliver a company of troops and a radar unit to Point Gorango, located on the northern coast of Morotai. The troops were to establish a radar station there. However, since the LCI gunboat no longer had her ramps for delivering troops, she instead towed an LCM behind her, which was loaded with the seventy army troops. It would take all night for the LCI gunboat, the LCM behind her, and the rest of the convoy around her to get there. At 6:25 a.m. on the morning of September 21, the convoy spotted enemy planes as they rendezvoused with the destroyer USS *John Rodgers*. The submarine chaser, PC 1122, which was also with the convoy, selected a landing area and all ships remained on standby as the *John Rodgers* shelled the beach. After several hours, *John Rodgers* ceased firing, at which time the LCI 227, LCI 70, and the 65 made one sweep at the beach firing all their guns. The 65 was the last ship in the column and it was reported that she drew some 20mm fire as they turned from the beach. At around 9:30 a.m., the 65 ceased all her fire. Several minutes later, the LCI (R) 73 began making its run into the beach, but after lining up and ranging their targets, the rockets would not fire due to a faulty circuit. The LCMs full of soldiers, including the one the 65 had been towing, then made their landings. Once again, luckily, those landings would go unopposed.

The LCI 65 and LCI 70 surveyed the landings off the beach, waiting on standby in case they were needed to lend support. Shortly after 11 a.m., the 65 joined the 70 in patrolling a nearby village down the beach several miles to the northwest but still on the northeast side of Morotai. After half an hour, the two gunboats found no enemies or signs of life present in or around the village and were soon ordered to return to the Cape Gila area of Morotai with the LCI 70 and PC 1122. Shortly after 7 p.m., Stephen and the crew of the 65 arrived off southwestern Morotai with various other landing craft and large ships.

The LCI 65 spent the next several days conducting routine patrols with various ships off the coasts of Morotai, Rao, and Halmahera Islands ("Able" area). On September 26, while the 65 was anchored off Cape Gila, Macaluso held a deck court for Stephen's buddy, Paul

DeWitte. Shortly before noon, Macaluso read the charges aloud which found DeWitte guilty of "sleeping on watch." Earlier that summer, Stephen had taken the fall for DeWitte when the men had been caught stealing beer at Florida Island. The officers had threatened to kick DeWitte out of the Navy with a dishonorable discharge. It was Stephen who took the charge that had saved DeWitte then. This time it was being on the front lines in a combat zone that saved DeWitte. He would get a demotion to the next inferior rating from Fireman 2/c to a Fireman 3/c, but would once again avoid a dishonorable discharge.

As the LCI 65 sat anchored off Morotai on the morning of September 27, the weary and exhausted crew desperately tried to get some sleep in their tiny bunks belowdecks after two weeks of nonstop patrols and action. However, at this point the young men understood that aboard their LCI a full night's rest was as far away as the dark side of the moon. Shortly after 6 a.m., the alarm blared for General Quarters, so the men jolted awake and rushed to their battle stations. Even the idea of regular bowel movements could make an LCI sailor jealous, as records would later show that Stephen had been suffering bouts of severe constipation. One can only imagine the misery of trying to relieve oneself while sitting on what was essentially just a wooden plank with a circular hole, while being tossed and bobbled around on the turbulent Pacific waters. It was a common occurrence to be called to General Quarters while "hitting the head." Even on the commode, the call to General Quarters would rip a sailor away, as he was expected to run to his battle stations as quickly as possible no matter what. After the "all clear" was given, the sleepy sailors returned to their bunks to try their best to sleep once again—or they scampered back to the head to try to finish relieving themselves, but usually by then, a sailor didn't have to go anymore.

Later that afternoon, the LCI 65 received orders from Commander Task Unit 77.3.27 to be towed by LST 204 all the way back to Hollandia, because the LCI 65 had lost her screw. Nonetheless, the 65 had completed her duties at Morotai and was headed back to New Guinea to prepare for the next big invasion: Leyte, Philippines.

According to Morison's history, the Allied capture of Morotai proved to be even more valuable than originally anticipated. The airfield on Morotai was the only Allied base in the area that could strike Leyte with short range fighter planes and short-to-medium range bombers, but it was also the only base from which long-range aircraft could scout north and west of Leyte.[11] Even after the invasion of Leyte, the bases on Morotai Island helped the Allies strike Mindanao and Borneo in the months to come. On top of that, Morotai also turned out to be a strategically valuable base for PT boats.

Just as the convoy of Allied ships had formed up and gotten underway for Hollandia that evening, the men spotted an enemy plane. The alarm blared for General Quarters and Stephen rushed to his battle station along with the rest of the crew. The Japanese plane dropped a bomb near the convoy. It is unknown what happened to the Japanese plane after that.

One of the most vivid memories Stephen had from Morotai was witnessing the first Japanese pilot attempt a suicide crash dive. Though there is overwhelming evidence and consensus among military historians that it was at Leyte Gulf, Philippines, that the Japanese first used suicidal kamikaze pilots as its main weapon, Stephen nevertheless insisted he saw a pilot attempt a crash dive more as a last act of desperation than as a primary tactic. Morotai was "where I saw the first kamikaze," Stephen said in 2011. "He was up there about fifteen thousand feet. And boy she took a nosedive and started down there I thought, man, when is this sucker going to pull out?" He motioned with his hands before adding, "and man he just, '*ffff-pheewwww*,' right alongside of [a freighter]." He continued, "When I was on that little island I was telling you about, Morotai, that was the first Jap suicide plane I'd seen." Stephen said he recalled thinking at the time, "Man that guy is stupider than shit. He can't pull that plane out of there [from a nosedive]. Well, he didn't pull it out." Stephen realized the pilot had committed suicide on purpose. If the Japanese pilot hadn't missed that freighter, Stephen said, "He'd've went through the ship." The shock of it was something he would remember until his dying day.

It was a terrifying new weapon that the Japanese would use in battles to come. In 1274 A.D. and again in 1281 A.D., Japan had been saved from the Mongol emperor-invader Kublai Khan by the sudden arrival of Pacific typhoons. Because of these events, a legend arose that the gods favored the Japanese and that they were protected by the "divine wind," or in Japanese: the *kamikaze*.[12]

16

OCTOBER 1944: LEYTE GULF, PHILIPPINES

"Every day the air-raid siren would blare, sending us into the 'graveyard in the sky.' It was only a matter of time before all of us were killed."

– Toshimitsu Imaizumi, Japanese Petty Officer 2/c[1]

"When we got through with this Morotai Island, it was just a preliminary shot for this," Stephen said as he motioned toward the southeastern Philippines on the map. "And then from there, we went into Leyte Gulf. And that was an adventure."

As he had been doing all afternoon that day in May in 2011, Stephen kept his eighty-six-year-old thumb in motion as it slid from Morotai Island to the Philippines on the map in front of him. "See this one, Battle of Leyte Gulf, it would become one of the largest sea battles ever fought. And it would be the first time the Japanese first used the kamikaze weapon as a weapon." Before recalling several detailed memories from the battle, he looked over the page in front of him in solemn silence for a long moment then started by saying, "I could tell you the story about the Leyte landing..."

Modified version of Philippines map: https://d-maps.com/carte.php?num_car=590&lang

The Philippine Islands are made up of more than seven thousand islands. According to *National Geographic*, the island country can be divided into three main areas: Luzon (the largest, northernmost island, which includes the capital, Manila); a group of islands called the Visayas (including the major islands Panay, Negros, Cebu, Bohol, Leyte, Samar, and Masbate); and Mindanao, the second-largest island in the Philippines, found at the southern end of the archipelago.[2] Many of those islands are less than one square mile in area, but among those more than seven thousand islands is the eighth-largest one, the island of Leyte.

Volumes of books have been written about the Battle of Leyte Gulf that have explored the massive amounts of forces involved on both sides—which was staggering. There are many different unique aspects and layers to the Battle of Leyte Gulf that began in October 1944, but for the purposes of Stephen's story, we'll need to focus mainly on three key aspects of the Battle of Leyte Gulf—which was made up of four different engagements that lasted over several days.

The first aspect we'll need to focus on for the purposes of Stephen's story is one that has already been touched upon: that the Battle of Leyte was the first time the Japanese, out of sheer desperation at this point in the war, unleashed deliberate kamikaze suicide plane attacks against Allied ships as a primary tactic in the war. Japanese pilots would take off in planes loaded with bombs and explosives, and once encountering American ships, would purposely crash-dive and steer their planes into the ones they picked out as targets. The Japanese high command was under the belief that this tactic ensured better odds of a hit, even though it meant certain death for their badly needed pilots. And in many cases, the high command was right; the tactic proved to be deadly effective. To Stephen, the notion of pilots intentionally committing suicide was as baffling as it was terrifying.

The second aspect is that the Battle of Leyte began as a land operation, but days later turned into what became, officially, the largest naval battle in history: the Battle of Leyte Gulf. For the Allies, it started on the morning of October 20, 1944 (dubbed A-Day), as an amphibious landing of General Walter Krueger's Sixth U.S. Army on the mainland of Leyte Island. Labeled as "Operation King Two," the amphibious landing operation was made up of two different landing forces that established two completely separate beachheads at two different places on Leyte Island's eastern coast. Those two landing forces consisted of the northern force at Palo near Tacloban Airfield (X Corps), and the southern force at Dulag Airfield (XXIV Corps). Stephen's LCI 65 would be part of the Northern Attack Force, which covered the landings at Red Beach near the Palo River south of Tacloban Airfield. As the LCI 65 patrolled Leyte Gulf off Tacloban Airfield in the days following the landings, four massive air and naval battles broke out between the Allied and Japanese fleets, thus beginning the Battle of Leyte Gulf. Those four naval engagements unfolded as follows: the Battle of the Sibuyan Sea, the Battle of Surigao Strait, the Battle off Cape Engaño, and the Battle off Samar. Stephen's story at Leyte is deeply affected by two of those engagements: the Battle of Surigao Strait (Oct. 24–25) and the Battle off Samar (Oct. 25).

The third aspect important to Stephen's story is another one already touched upon throughout this book, that the actual landing of ground forces on the island of Leyte—King Two—was General MacArthur's long-awaited return to the Philippines. No one had lobbied military leaders harder or pressed them more vociferously than MacArthur had for a return to the Philippines. Although original plans before mid-1944 weren't certain if an invasion in the Philippines would even take place, General MacArthur and Fleet Admiral William D. Leahy convinced President Roosevelt to allow for the liberation of the Philippines. In his history, Barbey remembered being present in Honolulu in July 1944 as MacArthur successfully made his case that the United States government had promised the Filipinos that they would "return" and be rescued by the Americans. When Admiral Barbey had waded ashore Morotai Island on D-Day,

General MacArthur broke the startling news to him that Leyte Island would in fact be their next major target and first objective in recapturing the entire Philippine Islands, and best of all, according to MacArthur, the invasion would occur sooner than previously anticipated. The general was jubilant. Years of arguing to get his way with President Roosevelt and the Joint Chiefs of Staff had finally paid off. MacArthur's dream of a triumphant return to the Philippines was about to come true. Instead of December, the invasion of Leyte Island was moved up to October 20, 1944. Barbey agreed to return to Hollandia as soon as possible to plan their next invasion.[3]

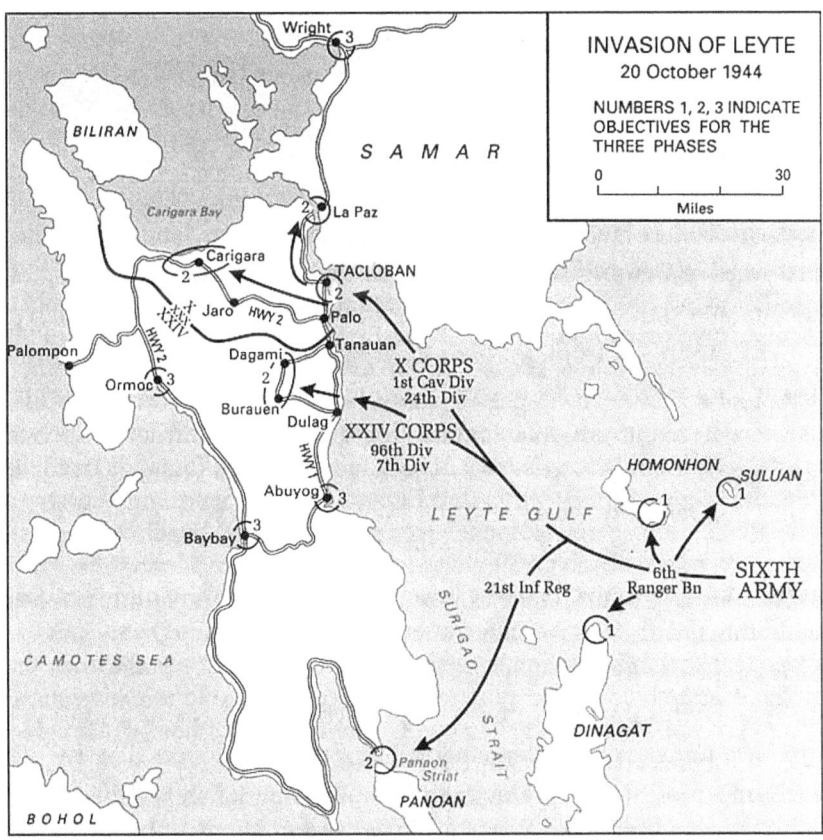

The invasion of Leyte Island on A-Day, October 20, 1944. From the official history, *U.S. Army Campaigns of World War II.*

October 1–12: . . . LCI (G) 65 was returning from MOROTAI to
HOLLANDIA in tow of LST 204 due to lost screw. She arrived on 2
October.[4]

Hollandia had become a major hub of activity in autumn 1944
according to Admiral Barbey. Army and Navy commanders from all
over the Pacific had flown in to assist in the planning of the Leyte
operation, and it seemed like every major commanding officer in the
Southwest Pacific had moved his headquarters to that area.[5]

On October 1, 1944, Stephen was promoted from Seaman 1/c to
Quartermaster 3/c. He was back at the rating he had before he got in
trouble for the beer-stealing incident when he took the fall for
DeWitte. Since the LCI 65 was still being towed by the LST 204, they
spent their time conducting gunnery drills. The next day, the 65
arrived back at Hollandia, New Guinea, still in tow. They spent the
first week of October resupplying and reorganizing in preparation for
repairs they'd need to undertake before the coming Leyte campaign.

On the morning of October 8, as she sat moored alongside the
repair ship USS Amycus, the 65 began pumping out her ballast tanks
as she entered an auxiliary repair dock for repairs on her propeller
shafts and rudder posts. Once she entered drydock shortly after 9
a.m., Stephen and the rest of the crew began to scrape the hull of
their landing craft, and eventually applied a fresh paint job to the
gunboat. "We painted it that camouflage color [so] we could blend in
with the terrain up in there once we got up next to the shore,"
Seaman 1/c J.R. Reid remembered.

Along with the 65, the rocket equipped LCI (R) 73 was also in the
drydock undergoing emergency repairs until the next day. In the
early morning hours, welders came aboard and finished installing
shields on the two forward 20mm gun mounts behind the square
conning tower. With her repairs completed, all the 65 needed now
was more fuel. She was out of drydock and water bound again shortly
before 10 a.m. on October 9. She pulled alongside the YOG-59, port

side to, and commenced taking on fuel. The crew spent the next several days gathering supplies and stores. They also took aboard several extra men for transportation to Leyte. On October 12, a giant pre-invasion rehearsal of the Allied naval forces took place off New Guinea in preparation for the coming battle. However, during the rehearsal the 65 encountered problems with her SCR-610 transmitter, so a radio technician from the LST 395 came aboard around 1:30 p.m. He quickly fixed the issue and was on his way back to his LST several minutes later. With everything else in working order, the 65 moored alongside the LCI (G) 23 and LCI (G) 70 in Jautefa Bay off the coast of Hollandia, where they'd remain for the night. According to Archie Holmes, the commander of LCI (G) Group Forty-Five, "On 12 October LCI (G) 23, 64, 65, 68, 69 and 70 were in all respects ready for sea and for the LEYTE operation."[6]

Stephen and the crew had no idea of knowing at the time, but this would be the last calm, peaceful night's sleep the men of the LCI 65 would be getting for the next four months. Instead of restful silence, the young men would soon become accustomed to the haunting General Quarters alarm that would blow, unexpectedly piercing the blackness in the middle of the night—swiftly ordering the tired, nervous men to battle stations against an enemy that always seemed to be everywhere.

————

Like Stephen, John Dingell had been too young to enlist when war broke out in December 1941 and had been waiting impatiently. As Dingell remembered it, "I was still too young, and there was no way my folks would let me go until I turned eighteen." When that long-anticipated day finally did arrive, he could wait no longer. Dingell wrote that he marched down to the U.S. Army recruiter on his eighteenth birthday, July 8, 1944, and "I told the recruiter I wanted to be in the infantry, and he said, 'Good, that's exactly what you're going to get.'" After several months, Dingell was finally on the war path himself. As he jokingly recalled in his book, "By the time the call finally came, on October 10, 1944, even my mother was about ready to see me get out of the house."[7]

However, just as soon as he shipped out to Camp Blanding, near Jacksonville, Florida, for his basic training, he got very sick from meningitis that he had caught from one of the cooks. Had it not been for the invention of penicillin, Dingell claimed that he surely would have died. It took him several months to recover, but soon after he did, he was sent to Fort Benning. As he remembered it, "Apparently, I'd impressed someone as a bright guy with leadership skills, and there was an opening at Infantry Officer Candidate School (OCS) at Fort Benning, Georgia. The army decided to try to make an officer and a gentleman out of me..."[8]

It was there at Fort Benning that Dingell earned the nickname "Artillery John." In his book, he humorously recalled the story that led to his namesake: "I was firing a 75-millimeter gun on the gunnery range and wiping out all the targets on the range. Wanting a bigger challenge, I shifted fire to the mortar range next door. Not long afterward, a very irate colonel appeared, demanding to know what was going on." Dingell said the colonel angrily shouted, "'Someone's taking out all my friggin' targets!'"[9]

––––––––

On the morning of Friday, October 13, fellow gunboat LCI 70 came alongside the LCI 65's starboard side. When the crews of both LCIs were mustered to quarters shortly after noon, the executive officer of the 65, William McKeon, gave brief instructions to his crew. Every single man, including Stephen, listened to McKeon as he spoke of the coming invasion, though the men still did not know their final destination. At 1:28 p.m. the 65 got underway in accordance with orders from Commander Task Group 78.1 for the coming secret operation number 101-44. By 2 p.m., the 65 had steamed out of Hollandia Bay and was speeding ahead to join the massive convoy that was the Seventh Amphibious Force's Northern Attack Force. It would take the convoy a week to reach Red and White Beaches at Leyte. The muggy humidity made for extremely uncomfortable living conditions for the men, especially when they had to wear life jackets.

USS LCI (G) 65
Crew list as of October 1944

Name	Rating/Rank	Date of Enlistment	Date First Received Aboard	Place of Enlistment
MACALUSO, Charles J., Commanding Officer	Lieutenant (junior grade)	N/A	N/A	South Bend, IN
McKEON, William J., Executive Officer	Ensign	N/A	11/8/1943	Lansing, MI
McGINNIS, John F., Gunnery Officer	Ensign	N/A	3/20/1944	San Francisco, CA
KINSINGER, Elmer H., Engineering Officer	Ensign	N/A	5/2/1944	Eddyville, IA
CLINE, Merle A., Officer Trainee	Ensign	N/A	5/17/1944	Hutchinson, KS
ABBOTT, Herman David	Signalman, 3rd Class	11/24/1942	7/25/1943	Columbia, SC
AISTON, Lester Eugene (KIA - 10/24/1944)	Ship's Cook, 3rd Class	1/26/1943	7/25/1943	Monroe, MI
BLIGH, Charles W.	Radarman, 3rd Class	7/20/1943	3/10/1944	Jerseyville, IL
BROWN, William V.	Radioman, 1st Class	2/14/1942	12/14/1942	Scranton, PA
CAIN, Gail Franklin	Motor Machinist's Mate, 2nd Class	3/25/1943	8/15/1943	Cheyenne, WY
CLARK, Steve Warren	Electrician's Mate, 3rd Class	12/8/1942	3/12/1944	Witchita Falls, TX
CRICHTON, Robert Galloway Jr.	Pharmacist's Mate, 2nd Class	11/2/1942	10/13/1944	Minneapolis, MN
DeWITTE, Paul Cyriel	Fireman, 3rd Class	12/30/1943	4/5/1944	Chicago, IL
DOLAN, William Henry	Motor Machinist's Mate, 3rd Class	7/12/1943	3/12/1944	Rochester, NY
EDWARDS, Gerald Leroy	Motor Machinist's Mate, 3rd Class	10/9/1942	3/12/1944	Duluth, MN
ERWIN, Harmon Eugene	Seaman, 1st Class (Cox)	10/9/1942	3/12/1944	Peoria, IL
FALEIDE, Donald James	Seaman, 2nd Class (MoMM)	1/21/1944	6/9/1944	Minneapolis, MN
FITZGERALD, Truman Page	Seaman, 1st Class (SC)	12/15/1942	3/12/1944	Shrevport, LA
GANZBERGER, Stephen	Quartermaster, 3rd Class	8/26/1942	1/16/1944	Detroit, MI
GIBSON, Fount (n)	Seaman, 1st Class	8/6/1943	3/12/1944	Barbourville, KY
GRECO, Arthur Anthony	Ship's Cook, 3rd Class	12/1/1942	2/24/1944	New York, NY
GREER, Robert Kenneth	Motor Machinist's Mate, 1st Class	10/26/1941	10/5/1943	Orlando, FL
GRIMES, Robert Claton	Gunner's Mate, 2nd Class	5/4/1942	3/18/1944	San Francisco, CA
HALL, Harry Melvin	Gunner's Mate, 2nd Class	5/4/1942	3/18/1944	Houston,TX
HEISEL, Michael Wilson	Motor Machinist's Mate, 3rd Class	12/3/1943	4/29/1944	Cincinnati, OH
HENSLE, Billy Earl	Seaman, 1st Class (Cox)	9/24/1942	4/25/1944	Los Angeles, CA
HOLLAND, Robert Louis	Seaman, 1st Class	7/21/1943	2/24/1944	Oklahoma City, OK
HOLMES, Roy Calvin	Seaman, 1st Class	12/13/1943	5/2/1944	San Francisco, CA
HONSTEIN, George	Seaman, 1st Class	10/19/1943	5/2/1944	Boise, ID
HOOK, Frank Jr.	Seaman, 1st Class	1/13/1944	5/2/1944	Kansas City, MO
HUDSON, Archer Bryant	Radioman, 3rd Class	8/2/1943	10/13/1944	New Haven, CT
JOHNSON, Odis Lee	Coxswain	11/28/1942	3/12/1944	Jackson, MS
KASPEREK, Edward (n)	Motor Machinist's Mate, 3rd Class	11/27/1943	4/29/1944	Detroit, MI
KING, Shellie Sr.	Steward's Mate, 1st Class	6/8/1943	2/8/1944	Jacksonville, FL
LADD, Willi Robert	Steward's Mate, 2nd Class	12/15/1943	4/16/1944	Houston,TX
LaBELLE, Glen Walter	Seaman, 1st Class	6/4/1943	4/27/1944	Minneapolis, MN
LEBER, Edward Robert	Seaman, 1st Class (RdM)	4/7/1943	4/27/1944	Rockford, IL
LIESKE, Arthur Francis	Seaman, 1st Class (EM)	6/29/1943	4/25/1944	New York, NY
MAHAN, Albert Wayne	Seaman, 1st Class	3/2/1943	4/16/1011	Indianapolis, IN
MARKHAM, Johnny Milbrun	Seaman, 2nd Class	9/22/1943	10/10/1944	Shreveport, LA
NIZNER, John Frank	Gunner's Mate, 2nd Class	4/15/1943	3/14/1944	Canton, OH
OBERSON, Stuart Grant	Seaman, 1st Class (GM)	3/1/1943	3/14/1944	Detroit, MI
ORBICH, Joseph Jerome Jr.	Seaman, 1st Class (QM)	2/27/1943	3/12/1944	Pittsburgh, PA
REEDER, Henry Hayman	Seaman, 1st Class	1/13/1941	3/12/1944	Huntington, WV
REESE, Clyde Robert	Seaman, 1st Class	12/28/1943	6/9/1944	Spartanburg, SC
REID, James Kent	Signalman, 3rd Class	12/14/1942	3/12/1944	Lynchsburg, VA
REID, J.R.	Seaman, 1st Class	12/21/1943	6/9/1944	Hattiesburg, MS
RUSSELL, Rex R.	Signalman, 2nd Class	2/10/1942	12/14/1942	Ottumwa, IA
THOMAS, Thomas	Seaman, 2nd Class	2/1/1944	10/10/1944	Cincinnati, OH
TOFT, Donald Richard	Gunner's Mate, 3rd Class	11/25/1942	1/27/1944	San Francisco, CA
VINCENT, James Oliver	Gunner's Mate, 2nd Class	2/8/1942	10/10/1944	Raleigh, NC
WOJCIECHOWSKI, Edward Francis	Boatswain's Mate, 2nd Class	8/27/1942	12/14/1942	Chicago, IL
YOUNG, Emory Shearld	Gunner's Mate, 3rd Class	11/21/1939	2/1/1944	Cape Girardeau, MO

Commanding the Seventh 'Phib's Northern Attack Force aboard the flagship USS *Blue Ridge* was Rear Admiral Barbey. After leaving Hollandia Harbor, the convoy of 119 ships assumed a modified circular formation and in a few hours every ship was where they were supposed to be and on their way. It wasn't until the next day, October 14, that Barbey informed the rest of the convoy over the public announcing system that they were in fact sailing toward the Philippines. Up until that point, only those who needed to know had been aware of the coming destination.

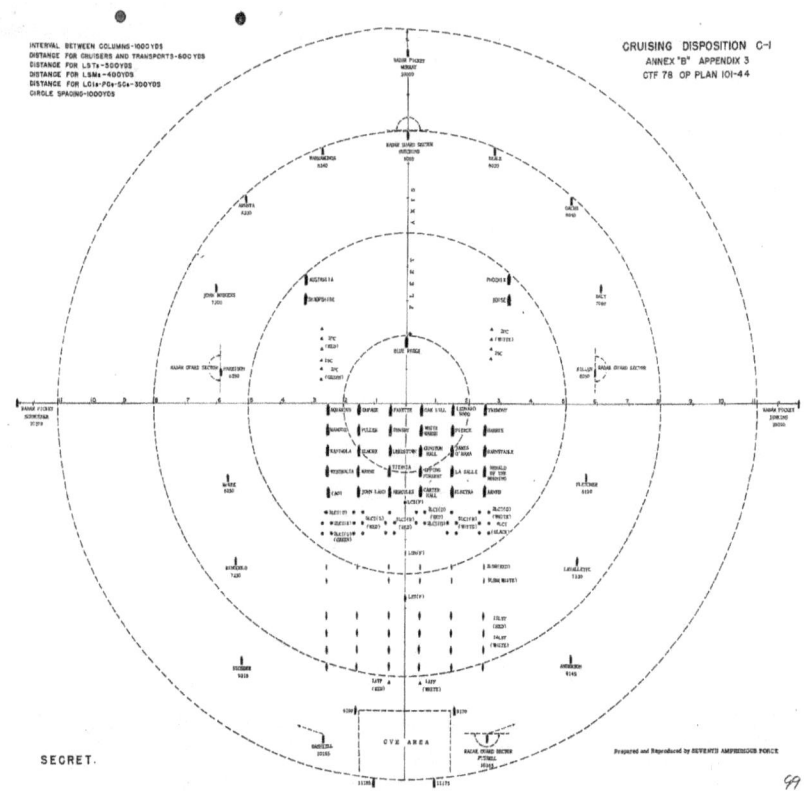

A diagram of the Allied ships' positions in Stephen's convoy on their journey to Leyte, Philippines. (*Action Report, Commander Task Unit 78 – Leyte Operation 1944*, p. 99)

The Southern Attack Force, which was not under Barbey's command, but rather Vice Admiral T. S. Wilkinson's command, would meet up with Stephen's convoy later, as the Southern Attack Force had left from Manus Island on October 14. The Southern

Attack Force fell under the overall command of Admiral Nimitz's Central Pacific Fleet, not MacArthur's. Both the Southern and Northern Attack Forces were scheduled to hit their designated beaches at Leyte at the same time on October 20.

Further instructions were briefly given to Stephen and the rest of the LCI 65's crew on the coming operations at Red Beach on October 18, two days before invasion day. Onward they steamed, under radio silence, toward Leyte Gulf as part of "Charlie One" cruising formation.

———

In the sticky pre-dawn darkness of the morning of A-Day, October 20, 1944, Stephen stood at the bridge sweating in his uncomfortable life jacket and helmet atop the LCI 65's square conning tower while Orbich, the other quartermaster, stood ready within the pilothouse below. Stephen could see Leyte Island in the distance ahead and hear the heavy, rumbling gunfire from the main battery of the battleship USS *West Virginia* as her shells unloaded on Red Beach. Soon, the sounds grew louder as the battleships USS *Mississippi* and *Maryland* joined the bombardment. As the quartermaster, Stephen remembered his duties that morning saying he had to "take care of all the charts, and make sure the ship was all set . . . navigation-wise for the captain and the navigating officers."

With the exception of several showers, the weather was clear and the sea calm. Communications were noted as good while using their recently repaired SCR-610 transmitter. The LCI 65 fell in astern (behind) the rocket ships LCI (R) 74 and LCI (R) 331 and headed toward Leyte's beaches. The cacophony of deafening firepower that quaked and shook and trembled the beach only grew louder as the light cruisers also opened fire. Stephen remembered the heavy shells' devastation adding, "When they hit the ground, the ground shakes for a half a mile around from the vibration."

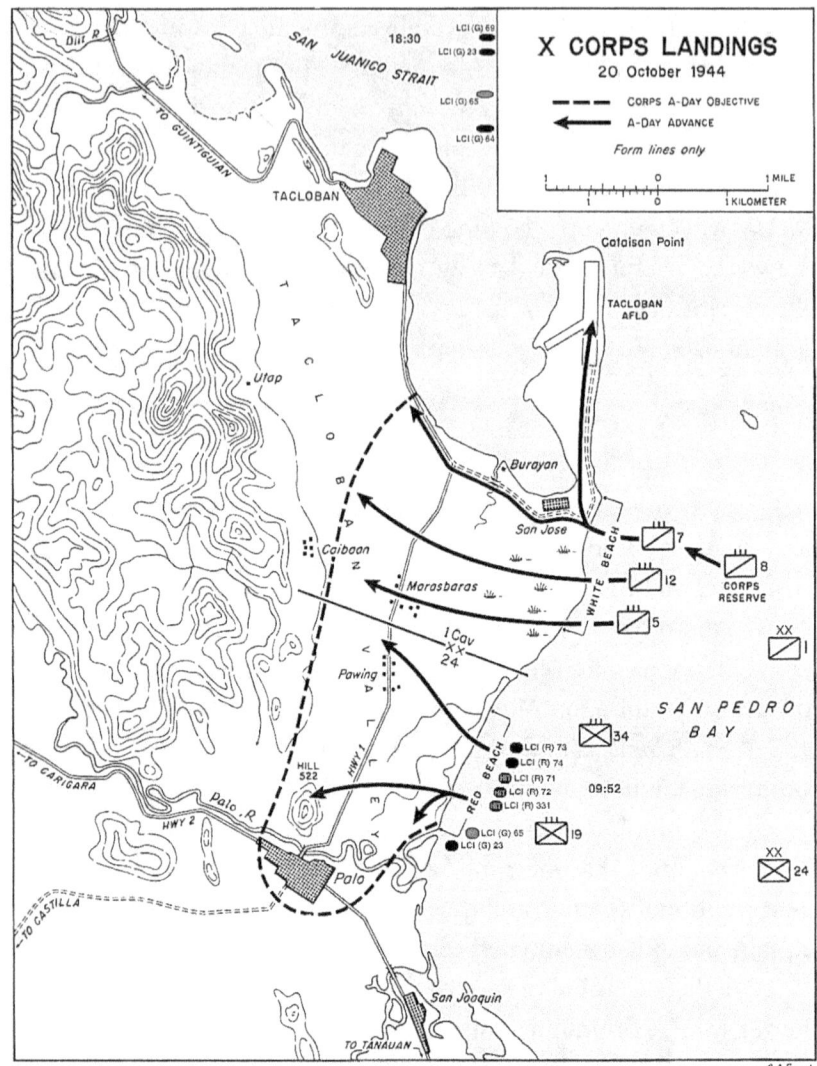

This map of the Northern Attack Force landing at Red and White Beaches
adapted with added labels from M. Hamlin Cannon's, *Leyte: The Return to the
Philippines, U.S. Army in WWII, The War in the Pacific* (1993). Note USS LCI (R) 71,
USS LCI (R) 72, and USS LCI (R) 331 were hit by Japanese 75mm fire as the first
waves approached and shortly after the LCI gunboats and rocket boats began
bombarding Red Beach around 9:52 AM. Also note several of the LCI gunboats
guarding the San Juanico Strait by 18:30 (6:30 p.m.), including Stephen's LCI (G)
65.

Stephen's gunboat was part of LCI support unit 78.1.8, which consisted of the following gunboats and rocket equipped LCIs as they approached the beach (left to right):[10]

USS LCI (G) 23
USS LCI (G) 65
USS LCI (R) 331
USS LCI (R) 72
USS LCI (R) 71 (Group Flagship)
USS LCI (R) 74
USS LCI (R) 73

The jobs of the LCIs were to cover the landings at Red and White Beaches between Tacloban Airfield and the Palo River. The LCI 65 and LCI 23 were assigned to cover Red Beach's left flank as the Army's 24th Infantry Division landed on Red Beach. At 9:45 a.m. the LCI 65 sailed in toward Red Beach ahead of the first waves of the landing craft filled with vehicles and troops (LCVPs). That's when the official orders broke through the radio:

> *"LCI(R) and LCI(G) Red Beach: proceed to assigned stations keeping to the westward of RED Beach transports."*[11]

In flank formation, Stephen remembered the exact moment when the bombardment from the battleships, cruisers, and destroyers ceased—and when his own gunboat opened up firing her three-inch 50 caliber heavy gun. He began, "Well, we'd stand off three hundred yards out, and if we picked up on any [enemy] firepower from the beach, then [our] three-inch rifle would get after them. And see, if we couldn't touch them, then the wagons (battleships)—we'd radio out to the wagons, or a cruiser." Stephen chuckled as he reflected on the enormous firepower of the battleships, "Oh man, their sixteen-inch [gun] was . . . you could see 'em coming." His hand imitated a shell moving in an arching motion. "Slugs, coming through the air."

That crucial moment for which only the rocket equipped LCIs and gunboats could be depended on had arrived again. At 9:52 a.m., the shelling of Red Beach had fully commenced by the LCI gunboats. Together, the LCI 65 and LCI 23 (on the 65's port side) shelled the left flank of Red Beach near the Palo River with every gun they had. As the two gunboats protected their left flank, the LCI (R)s 71, 72, 73, 74, and 331 began blasting Red Beach with their rocket barrage, eviscerating everything in their path. The troops of the 19th and 34th Regiments of the 24th Infantry Division watched as their LCVPs approached the scorched beach that now lay ahead of them. As thousands of LCI rockets hit the beach with a giant roar, the ground once again rumbled and shook. Official reports from that morning stated, "This rocket barrage was the best and most effective observed to date in the Southwest Pacific Area."[12]

And that's when it happened. According to official records, of the seven LCIs that made up Stephen's task force that day, three rocket ships were hit by Japanese 75mm shells. LCI (R) 71 had just unleashed her first salvos of rockets covering an area four hundred yards across the center portion of the beach by eight hundred yards deep past the water line, when she took a direct hit on her starboard side. Eight of her rocket launchers were knocked out, their electrical connections completely severed. Then, a few minutes later, as she strafed the left flank of Red Beach, the LCI 71's pilothouse was struck again by a 75mm shell that entered the starboard side and passed out the port side.[13] Nearby, fellow rocket ship LCI (R) 72 was not fairing much better. She had taken two hits from Japanese batteries, damaging her port side hull and the pilot house, but nonetheless her valiant crew somehow found a way to complete her mission of firing her rockets on Red Beach like bazookas. The LCI 72 also suffered eight casualties during their attack run. On the LCI 72's port side, the LCI (R) 331 had also taken a hit from a Japanese three-inch shell on her port bulwark, causing minor damage and wounding one man. Ironically, many of the LCIs at Leyte were saved from serious damage by their own light construction. Being designed as a troop transport and not a warship, enemy artillery shells frequently passed right through their thin steel skins, luckily, without exploding.[14]

After several minutes of bombardment, the LCI 65 and the LCI 23 executed a nine-turn (turning to port) as soon as the LCVPs passed them. The 65 continued strafing and shelling the beach from their starboard side with their three-inch 50 caliber, 40mm, 20mm, and .50 caliber guns. After hitting the beach with thousands of rounds of ammo, the 65 headed away from the beach and awaited further orders.

The three rocket LCIs in Stephen's task unit weren't the only ships hit. Shortly after 10 a.m., after the 24th Infantry and 1st Cavalry Divisions landed on Red and White beaches, respectively, two landing craft were shot up on White Beach and one sunk. According to records, three LSTs were enfiladed on Red Beach by enemy 75mm guns, which resulted in the death of one ship's skipper and a colonel.[15]

At 11:40 a.m., the LCI 65 received orders to go into the beach and give firefighting support to an LST that was on fire, thus beginning what Barbey described in his history as one of the two primary purposes of LCIs during the Leyte invasion: firefighting. Barbey stated, "The entire fire-fighting capability ashore depended on a few small LCIs."[16] The second primary purpose will be discussed a bit later.

However, after about ten minutes, the LCI 65 returned from the beach, because the LST no longer needed help and had retracted from the beach on its own. The 65 again stood by for the rest of the morning and afternoon and awaited further orders.

As they waited, Stephen caught a glimpse of General Douglas MacArthur making his way ashore on Leyte Island later that afternoon. Years later, Stephen remembered the exact spot as if it was yesterday, "That's where MacArthur walked on the beach," he said as he pointed to the map. But he wasn't a big fan of MacArthur's. Stephen cracked a joke about MacArthur's tendency to walk around smoking a corn cob pipe before adding, "Yeah, I saw him, . . . He didn't impress me." He explained, "I didn't like him because he left

[Lieutenant General Jonathan M. Wainwright IV] holding the bag, and the poor guy never been the same." Stephen was referring to the disaster Wainwright inherited back in March 1942 which had resulted from General MacArthur's secret evacuation from the Philippines and the surrender of American forces left behind. After war broke out and MacArthur escaped to Australia, it was General Wainwright who was bequeathed with the command of remaining Allied forces in the Philippines who were starving and outnumbered. By May 6, Wainwright was forced to surrender over seventy thousand American and Filipino troops, in what became the largest surrender of military forces in American history. The American prisoners, including Wainwright, were then—starving and dehydrated— brutally forced by the Japanese to march sixty-six miles, in what infamously became known as the Bataan Death March. Stephen remembered "They marched [the Americans] from Manila Harbor . . . If you looked crippled, [the Japanese] bayoneted you." Wainwright was subsequently the highest-ranking American prisoner of war (POW), and, despite his rank, was subjected to the same savage treatment by his captors. By the end of the war, Wainwright, who earned his nickname "Skinny," had become emaciated and malnourished from three years of torture and starvation in captivity. He had survived and earned the respect of all who were imprisoned with him, and he was eventually awarded the Medal of Honor in 1945 for his heroics in the Philippines. But Stephen found the idea of abandoning his own men, especially "fighting men" like Wainwright, as unacceptable and painful.

———

Shortly after finishing a luncheon in his cabin, MacArthur gathered himself on the deck of the light cruiser USS *Nashville* with a few members of his staff. He had spent that morning of A-Day observing the landings from the *Nashville's* bridge, but the symbolism of the moment he experienced must have been thrilling. MacArthur's long-awaited return to the Philippines was finally happening. Press correspondents who were present for the historic moment later described the scene.

General Douglas MacArthur and members of his staff land on Red Beach, Leyte, on October 20, 1944. (Photo by U.S. Army Signal Corps officer Gaetano Faillace, U.S. National Archives photo No. 531424)

Wearing freshly-pressed suntans and sunglasses, MacArthur made his way down a ladder into a small landing barge. Among those present with him were Philippine President Sergio Osmeña, Air Commander George C. Kenney, and Lieutenant General Richard K. Sutherland. As they approached Red Beach in the third assault wave, MacArthur could smell the burning palms. He turned to Sutherland (his Chief of Staff) and with a wide smile said, "Well, believe it or not, we're here."

When the landing barge got about fifty yards from the beach, the coxswain dropped the ramp. MacArthur stepped coolly into knee-deep water with the others and waded ashore. They studied the damage done by the massive naval bombardment, including the damage done by the LCIs. Then, standing triumphantly on the beach, MacArthur's voice took on a tone of deep emotion as he began his famous liberation speech:

> People of the Philippines: I have returned. By the grace of Almighty God our forces stand again on Philippine soil . . . At my side is your president, Sergio Osmeña, worthy successor of that great patriot Manuel Quezon, with members of his Cabinet.

But the fervor in MacArthur's voice grew progressively richer and more absolute as his speech went on:

> Rally to me. Let the indomitable spirit of Bataan and Corregidor lead on. As the lines of battle roll forward to bring you within the zone of operations, arise and strike! . . . For your homes and hearths, strike! For future generations of your sons and daughters, strike! In the name of your sacred dead, strike! Let no heart be faint. Let every arm be steeled. The guidance of divine God points the way. Following in His name to the Holy Grail of righteous victory!

At the conclusion of his speech, he sat down under the shade of a palm and briefly spoke with president Osmeña. He scribbled a short letter to President Roosevelt informing him of the successful landing. Then he departed and began heading back to the *Nashville*.

It had been more than two years since MacArthur reluctantly fled the Philippines during a heartbreaking defeat to the Japanese in early 1942 at Bataan. His senior officers that he'd gathered together claimed that he initially refused to leave even after they insisted he do so. He only consented to leave after receiving a direct order from President Roosevelt. The truth was MacArthur loved his men. And before his evacuation to Australia he told General Wainwright to hold on until he came back. In Australia he proclaimed "I shall return." From that moment on, he had a burning resolve. And at last, here he was, once again standing on Philippine soil with an army at his back. He had kept his promise. Douglas MacArthur had finally returned.[17]

———

By the early evening of A-Day, Stephen's LCI 65 had gotten underway and proceeded to the northernmost part of Leyte Gulf with the LCI (G)s 23, 64, and 69, to San Pedro Bay—another miniature bay that forms a pocket between Leyte and Samar Islands at the mouth of San Juanico Strait. For the remainder of the night, the LCI gunboats anchored in the waters off Tacloban near the airfield on Cataisan Point. There, they'd begin their new primary objective for the next several days: guarding the San Juanico Strait and intercepting any Japanese torpedo boats or barges that might try sneaking through the much narrower body of water that separates Leyte and Samar. The four gunboats guarded the strait all night, but with the exception of numerous air raids, the patrols were negative—no enemy ships sighted.

As A-Day drew to a close, Tacloban airstrip had been captured and the 1st Cavalry had made the most progress because White Beach had encountered the least amount of enemy resistance during the amphibious landings. The forces that landed at Red Beach had taken a beating from Japanese artillery fire, but overall, the Allied landings were so successful because one of the most obvious things missing from the day's invasion had been Japanese air attacks. Convinced the Japanese had made a fatal error, high officials thought they had frittered away their air power. The reality was that the Allies had no idea what the Japanese air force was about to unleash on them.

The early morning hours of October 21, (A+1 day)[18] began under the eerie glow of star shells, as the LCI 65 went to condition red as enemy air raids began and made all preparations to get underway. She would spend the day constantly on the move, shifting to different spots within San Pedro Bay, making her way in a southeasterly direction, before making her way back up to the mouth of San Juanico Strait by days end. In that time, the LCI (G) 64 came alongside the 65 in order to transfer off a Steward's Mate 1/c by the name of H. Whitney that the 65 had been temporarily carrying for passage. After the 64 cast off with her new crewman, the LCI (G) 34 came alongside the 65 to do the same thing, except for a second-class gunner's mate named L. Hautlzhauer. With fewer passengers, Stephen's gunboat received orders to make their way back up to the San Juanico Strait just as the sun was setting. By nightfall, the 65 had joined fellow LCI gunboats 23, 64, and 69 for guard duty.

Stephen did remember one event from that day that stuck out in his mind. Earlier that morning, the heavy cruiser HMAS *Australia* was attacked by three Japanese planes. Two of them were shot down, but in the chaotic scene, the third managed to crash dive into the *Australia*'s foremast, making it the first kamikaze of the Leyte operation. Stephen recalled seeing the result saying, "one dove into the Australian cruiser, killed the admiral, and killed the staff that was up in the bridge." Stephen was referring to Captain Emile Dechaineux, described by his own men as an outstanding person, and the *Australia*'s navigator, Commander John Rayment, who were both killed in the attack. Among the wounded was the commander of the Australian Squadron, Commodore John Collins, who had been on the bridge when the petrol tanks exploded. In all, thirty men were killed or died of wounds, with sixty-four wounded—twenty-six of these seriously. Many of those men from the *Australia* were so badly burnt they were unrecognizable.[19]

The dawn began with sounds of artillery on October 22 (A+2 day), as Stephen and the crew of the 65 went to battle stations. In the distance, fires could be seen burning on Red Beach. After shifting positions, the crew witnessed the rest of the ships in the harbor open up on Japanese planes. At 8:50 a.m., while underway off Samar, the crew sighted a headless body floating in the water but was unable to identify much of anything. The 65 sent a signal request to the nearby destroyer, USS *Russell*, to investigate the matter further. By 9:45 a.m., Stephen's ship was on its way to the LST 455, which had recently been converted to a repair ship and renamed USS *Achilles* (ARL-41), because the 65 needed repairs to their starboard screw and shaft. After mooring alongside the *Achilles*, a diver from their ship went down underwater to check the 65's propeller. Within ten minutes the diver was back up reporting that one blade was missing from the 65's starboard screw, which would require drydocking in order to repair. The diver also reported other minor repairs would be needed. Shortly after working parties began bringing badly needed supplies aboard the 65, the gunboat's coxswain, Lewis A. Wright, was transferred to the USS *Blue Ridge* for medical treatment later that afternoon. After casting off from the *Achilles* in the late afternoon, the 65 picked a new anchorage and remained there for the remainder of the night. Stephen and the crew were called to General Quarters twice throughout the night in response to Japanese air raids. Hunger, sleep deprivation, constipation, and frayed nerves caused uneasy men. As he floated within the darkness of the pilothouse within the conning tower, one can only imagine the thoughts that must have been going through Stephen's mind that night. One can reasonably assume thoughts about whether he'd even make it back home were among those filling his head. But he may have been comforted by the thought that, unbeknownst to him, he would live to see the end of the war and would get to celebrate the birth of his youngest granddaughter in exactly fifty years to the day.

Famous *Chicago Tribune* war correspondent, Walter Simmons, once began a story, "The dawn comes up like thunder every morning and this is how it goes," when describing the daily life of American soldiers on Leyte Island. He continued: "Suddenly there is a sound like a giant hand beating a carpet. 'Whomp, whomp, whomp' it goes. It is a 40 mm gun battery signaling a raid alert. Soldiers and civilians leave their beds."[20] This was how the crew of the LCI 65 started their morning on October 23, a day that would also include their own interactions with several famous war correspondents. All men hurried to their battle stations to the distant sounds of 40mm ack-ack fire just as the sun was rising. Shortly after 8 a.m., Colonel J. A. Kemp of the British Army and Commander A. Ford of the Royal Navy, came aboard the 65 for transportation to the Dulag area on the east coast of Leyte. The 65 then picked up the following group of war correspondents from various publications for transportation to Red Beach:

- Martin Sheridan (*Boston Globe*)
- Sam Blumenfeld (*Midpacifican*)
- E. T. White (Associated Press)
- Lesle Shoemaker (United Press)

It is believed that these correspondents were attempting to make it over to Tacloban City Hall to get a glimpse of the action where one of the great moments in history was unfolding. That afternoon, Philippine President Osmeña and General MacArthur stood on the front steps to proclaim civil government had been reestablished in the Philippines.

The trip took about two hours and shortly before sunset Colonel Kemp and Commander Ford disembarked on small boats upon reaching their destination. Shortly after, the four war correspondents also departed the LCI in a small boat that pulled alongside. After bidding the correspondents farewell and good luck, the 65 made their way back up to their patrol area at the mouth of the San Juanico Strait. As they were underway, the alarm for General Quarters began blaring, so the crew hustled to their battle stations. Atop the conning tower, Stephen watched the 40mm gun unleash a torrent of over forty rounds on a Japanese airplane overhead. The day had ended with the same sounds that it had started with: the "whomp, whomp, whomp" of 40mm gunfire.

Every once in a while, while the men were waiting nervously at their battle stations in the darkness, Lester Eugene Aiston, or whoever was the cook on duty, would sneak food and drinks to the men at their guns, since the crew had to remain at their guns as long as the men were at General Quarters. Stephen remembered how the 65's beloved cooks, including Ship's Cook 3/c Aiston, used to bring the men food during the night, even into the wee hours of the morning, to quiet rumbling stomachs. Another veteran, signalman Russell Hartwell of the LCI (L) 711 recalled in his memoirs that their cook used to do the same for their crew during the Philippines campaign, writing, "For the rest of that day and night we slept on the deck at our battle stations between attacks. The cook and his helpers brought coffee and sandwiches up to us."[21] Those types of gestures meant everything to the crew. The cooks were among the most sacrosanct of all. Aiston was not only a good cook, he was also a good friend. That's why nothing could prepare Stephen or any of his shipmates aboard the 65 for what was about to happen the next morning.

―――――

"HERE HE COMES!" Stephen shouted from the conning tower the moment he saw the kamikaze headed for his ship.

An elderly Stephen recalled the story from that October morning, "I was watching through the binoculars, and at the time I was steering the ship," he explained, "and I kept watching him." He lifted his hands up to his eyes imitating binoculars and continued, "I kept looking at him like this, and pretty soon you could see that airplane was coming exactly where grandpa was standing."

Stephen remembered the day of October 24, 1944, more than any other day in World War II. In his words that was the day "when the Japs really came after us." Shortly before 7 a.m. on A+4 day, the LCI 65 had just changed positions and anchored in a new spot in San Pedro Bay at the mouth of the San Juanico Strait. At exactly 8 a.m., the following position was recorded in the 65's deck log:

11° 15' 30" N
125° 00' 00" E

Ten minutes later, the alarm whistled for General Quarters, as Japanese planes were spotted in the southern sky. Fighters began engaging in dog fights with Japanese planes. Upon seeing the action, the LCI gunboat quickly fired up her engines to get moving, but just as the crew was about to get underway, the crew spotted an additional twenty-five Japanese twin-engine bombers that appeared from the west over Leyte. At 8:13 a.m., the LCI 65's deck log noted that Allied planes attacked and shot down three enemy aircraft. At that moment, Engineering Officer Elmer Kinsinger, who had joined the 65's crew in

May 1944 wrote, "I remember thinking this was just like the movies."[22]

The Japanese bombers began to individually break off from the formation and as Stephen put it, "Each of them picked out a ship and dove into a ship." One of the bombers left the formation and came flying toward the 65 on the ship's port side. As the bomber approached the 65, the crew opened up fire vomiting bullets up into the sky. The three-inch 50 caliber and 40mm guns barked gunfire as well as the 20mm and the .50 caliber guns. Seaman 1/c J.R. Reid was the loader for gunner and Coxswain Odis Johnson on the ship's number three 20mm gun located at the back of the ship on the port (left) side. When asked about the incident, Reid remembered, "We had a [three-inch 50 caliber] that you could set the fuse on the [round] for it to explode so far out . . . and they shot [the bomber] three times, and [the bomber] decided he'd come over and take care of the ship." Beside Reid's and Johnson's gun was Ship's Cook 3/c Lester Aiston, acting as the gunner on the number four 20mm gun on the starboard side at the back of the ship.

The men's excellent shooting scored a number of hits on the enemy bomber, as it was still a distance away. But the bomber did not fall. Instead, the bomber began to strafe the ship. Kinsinger remembered, "Then it suddenly turned very deadly..." The bomber dropped what looked like a bomb in the water some distance away from the 65. But it wasn't a bomb. Once it hit the water it began to create a wake, moving directly toward the 65. Kinsinger added, "one of the twin engine bombers . . . came at us releasing his torpedo that passed under our ship."[23]

"I could see our three-inch rifle shaking him," Stephen motioned with his hands imitating a gun as he spoke. "Our three-inch rifle was going, '*pee-choo, pee-choo, pee-choo!*' And you could see it, trying to get that plane." Bullets from the three-inch 50 caliber heavy machine gun at the front of the ship, as well as bullets from Reid's and Aiston's 20mm guns at the back of the ship, scored direct hits which tore off a chunk of the bomber's left (port) wing and fantail. But still the bomber did not fall. "And then all of a sudden, he turns like this,"

Stephen curved his right hand and moved his arm counterclockwise, "and starts coming down the mountain, I go, 'Oh man here he comes!'" Stephen continued as he pulled his hands apart, "and . . . the doors open like this and '*w-hoooo!*' All of a sudden, he drops his torpedo. I thought it was a bomb! I thought, man, is that guy stupid or what? I go, 'Why the hell's he dropping a bomb way back there?' I didn't think of a torpedo, you know?" But in a wicked stroke of luck, because of the LCI's flat bottom, the torpedo passed right under the ship's stern, barely missing it.

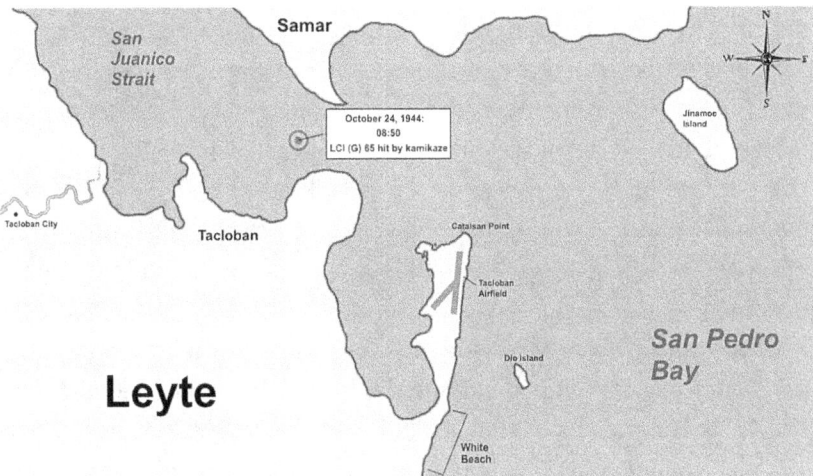

A map of the kamikaze attack on USS LCI (G) 65 on October 24, 1944.

But that's when the bomber began to fall. As it fell out of the sky, it became apparent it was on a collision course with the LCI 65's conning tower at the middle of the ship. This would've been a mortal blow to the ship and everyone on it. "I look over to [Macaluso], and man that guy was petrified he couldn't even talk. So, I just hollered down [the voice tube] real fast, 'all engines ahead flank,'" Stephen recalled. He indicated that by doing this, the ship began to shift its position to try to avoid collision. However, despite Stephen's best efforts to avoid impact, the Japanese twin engine bomber—smoldering in flames as it screamed downward—smashed into the back fantail of the 65 killing Stephen's close buddy, Lester Aiston.

"[The bomber] just caught the tail end of our ship," Stephen remembered somberly. "He killed one guy from Ida, Michigan. A young fellow by the name of Aiston." Stephen spelled his name out loud, and said, "he was a ship's cook too." Then for the first and only time that afternoon, Stephen choked up just a bit before adding, "He was a good kid."

J.R. Reid agreed, later saying, "He seemed like a mighty nice guy, to have to lose his life like that." Reid, on the port side 20mm next to Aiston's gun, remembered the moment Aiston was killed saying, "He was manning a twenty-millimeter on the starboard side on the stern, and . . . I was on the port side . . . [the bomber] came down on the stern and it just wiped the starboard side out over there with Aiston on it." Reid also remembered, "It was a Betty bomber, it wasn't a fighter plane." When asked if he remembered what Aiston looked like, Reid replied, "He was a big, tall guy, kind of heavyset. He probably [was] about six-foot, six-foot-one, and pretty well built." It all happened so quickly that Reid added, "Aiston never did know what hit him."[24]

"I am sure he was following orders to watch for planes and never saw the bomber coming at us from the bow," Kinsinger recalled. "The rest of us hit the deck. Had he done so, his life might have been spared." Kinsinger added, "I remember him as being a very nice [19] year old."[25]

"That guy just killed Aiston!" Stephen remembered hollering down from the conning tower, alerting his shipmates. "I got so excited I wanted to jump in the water and stab that sucker," Stephen said about seeing the Japanese pilot and airplane wreckage scattered in the water. "That's how mad you get."

Much damage was done to the LCI 65 in the attack. The stern fantail was damaged, including the stern winch, number three and four 20mm guns at the back of the ship, splinter screens, and stanchions.[26] J.R. Reid recalled the damage the kamikaze caused saying, "he . . . clipped stuff on top of the deck, he got the engine we used to pick the anchor up with, and the gun turret on the starboard side."

A sketch of the October 24, 1944 kamikaze incident drawn by the captain of the LCI 64 in his action report. Notice another kamikaze nearly missed the LCI 64 nearby. (USS LCI 64, *Action Report – Central Philippines – Leyte Operation*, dated November 1, 1944, p. 4)

The crew of the 65 quickly scrambled to regain control of the ship and assess the damage, as the kamikaze attacks continued on other ships around them in Leyte Gulf. Another kamikaze barely missed hitting the LCI 64, which was next to the 65. Several hours later, at 2:17 p.m., a small boat from LCI (D) 228 came alongside to assist with the collection of Aiston's body and dispose of it on the LST 464—which was acting as one of the Allies' only hospital ships that they had available at Leyte. The LST 464 had become the most important medical facility afloat or ashore, because hospital ships had been forbidden to enter the Gulf for the first few days because of the kamikaze threat. Because the Japanese were purposely targeting hospital ships, the LST 464 could only enter long enough to pick up casualties and get out.[27]

Two minutes later, with LCI 228 assisting with Aiston's body, the crew of the LCI 65—who had grown fond of their jolly, beloved cook—saluted Aiston and bid farewell to their buddy for the last time. According to the official action report of the LST 464:

At 1440 U.S.S. LCI(G) 65 came alongside with casualties and dead.[28]

Engineering Officer Kinsinger wrote, "I remember very clearly taking the body ashore where we met an Army burial crew. The sergeant told us to just lay him down and said they would get to him later. It was the most empty feeling of my life. Somehow I thought he deserved something more than that kind of treatment."[29] Aiston was now the 65's second man killed in combat.[30]

USS LCI (L) 1065 on fire and just before sinking in Leyte Gulf after being struck by a suicide Betty bomber on October 24, 1944. LCI 1065 was struck at the same time as LCI 65. This view is from USS LCI (L) 227. (Courtesy Brian Miller; originally from the Frank S. Ford estate)

Nearby, LCI (L) 1065 was also struck by a kamikaze Betty bomber. Witnesses described seeing the LCI bursts into flames in the distance. Burning oil from the LCI 1065 caused flames to shoot up 200 feet and billowing smoke to rise 5,000 feet high. Commander L. R. McDowell was watching from the *Blue Ridge*. In his war diary he wrote:

08:49: U.S.S. LCI burning[31]

By early evening, the LCI 65 anchored in San Pedro Bay off small Dio Island with several other ships. Even though she was four days into the invasion, and had just lost a man, the action was just getting started. And not just for the small LCI gunboat, but for the entire Allied navy in the Philippines. At 5:25 p.m., the gunboat went to condition red at General Quarters and made all preps to get underway to conduct the second primary job of LCIs at Leyte Gulf: smoke screening.

It was at Leyte that the Allies learned the real value of smoke laying LCIs. All too often, Japanese suicide planes would head for the biggest ship, break through the ack-ack anti-aircraft fire, and explode on the deck of their target. So the Allies were forced to deploy a different tactic: hiding. This was done by concentrating the larger, anchored ships in the same area, then LCIs would encircle them while laying a heavy pall of smoke all around. In doing this Barbey wrote, "The enveloping smoke saved many a ship."[32]

Lloyd B. Nothern, who served aboard the LCI (L) 612 as part of the LCI (L) Flotilla Fifteen staff, remembered the dangerous and dreaded smoke screen duty in the Philippines writing years later, "Many LCIs were underway every night in port 'laying smoke' in an assigned area. To prevent smoke dissipation, they maintained station at rather close intervals so that smoke laid by the first did not evaporate until replenished by the next. While this greatly reduced suicide attack losses, it was a hated duty. At a time when we had no water for showers, that smoke generator drank several hundred gallons of water every night."[33] And to top it off, each Besler Generator only had enough oil to make smoke for about four hours.

But keeping the smoke from dissipating was made much harder on an LCI. With all the dangers and hazards already associated with smoke laying in a war zone with suicidal planes crashing down, the 65 also had to deal with being at a disadvantage when it rained because of their type of smoke layer when trying to hide or screen transports. Official reports warned in their smoke screening plan for the Philippines: "Scattered rain squalls occur almost daily with showers . . . usually in the afternoon and early evening . . . Rain reacts unfavorably to smoke generated from BESLER units and smoke pots, tending to wash it out."[34]

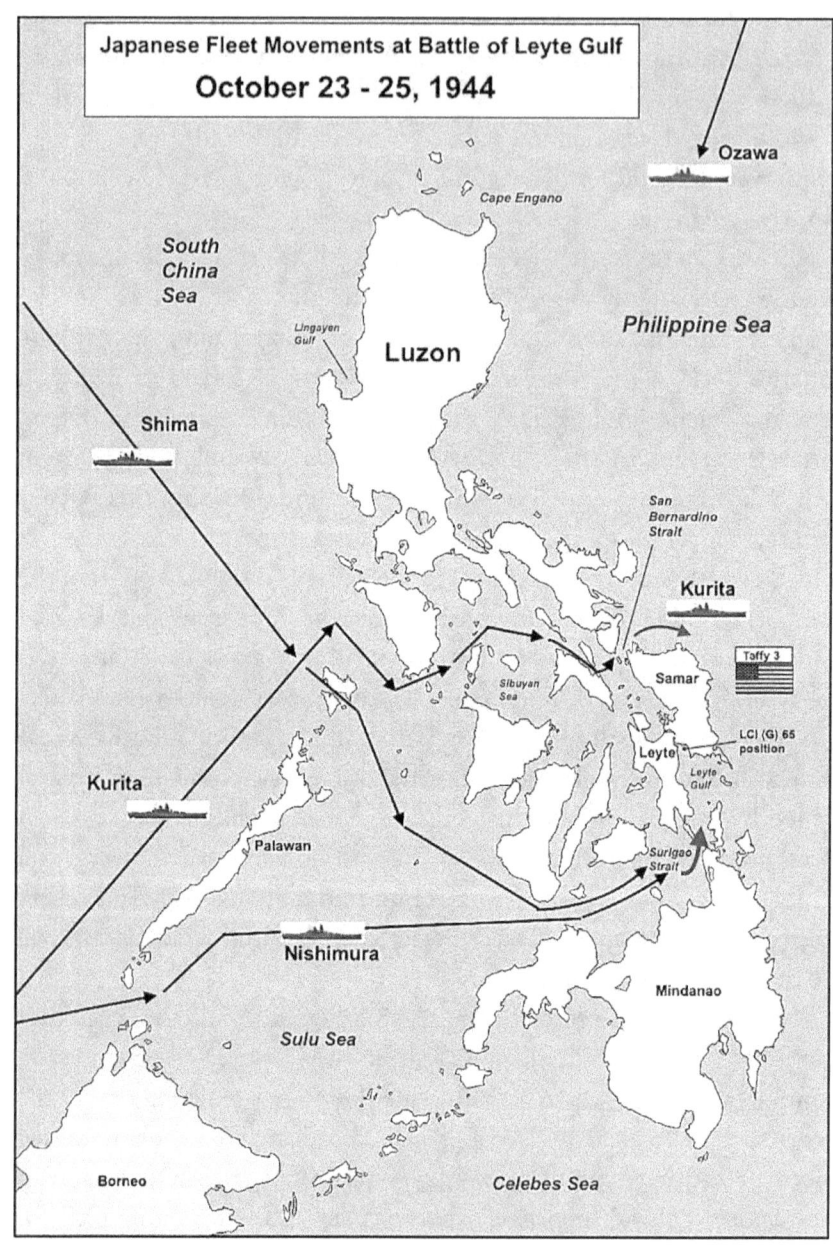

Modified version of original map: https://d-maps.com/carte.
php?num_car=590&lang=en

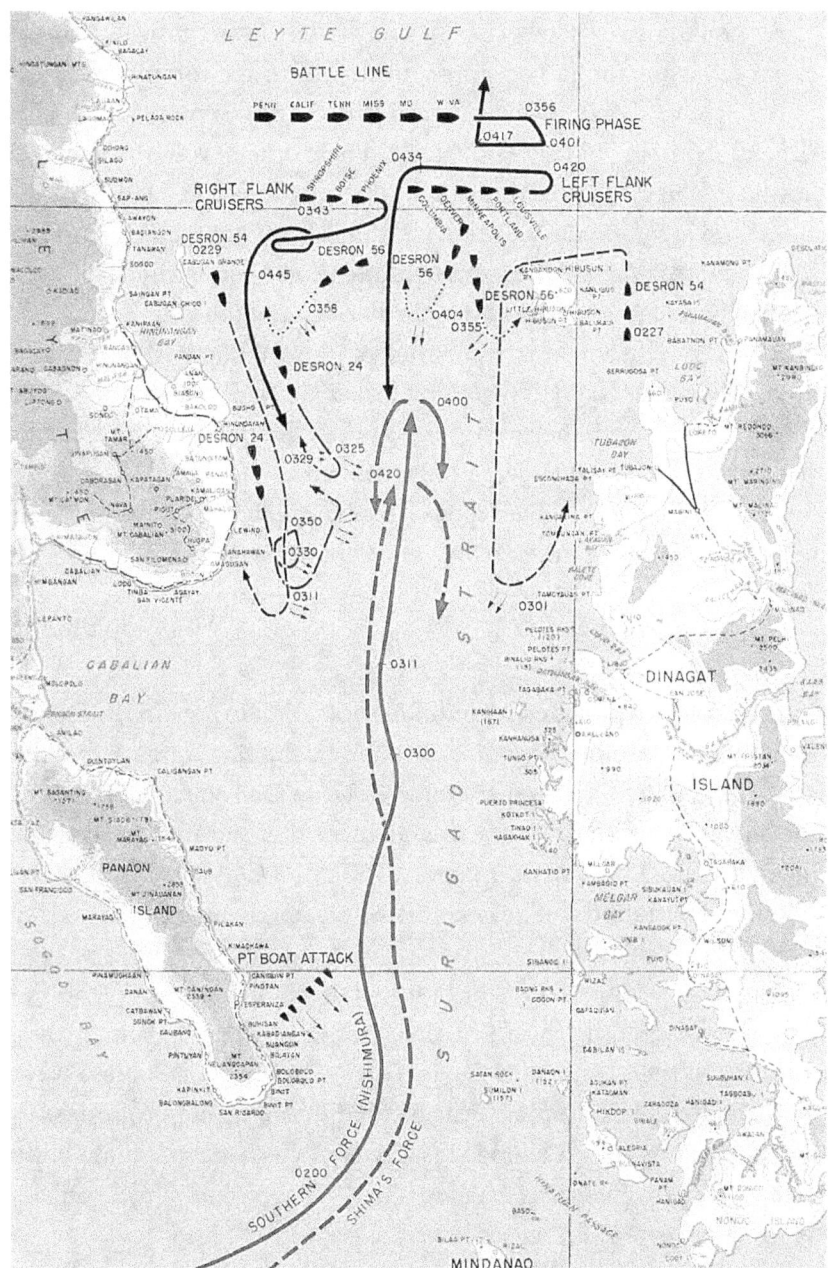

The Battle of Surigao Strait, early morning of October 25, 1944. (Reports of General MacArthur, 1994 facsimile printing, Vol I. Plate No. 61)

Several minutes later, the 65 was underway to provide smoke screen cover for the larger ships in the gulf. After completing their smoke screen duties, the gunboat settled in a spot off Tacloban Airfield just north of tiny Dio Island. Their orders were the same as they had been for the previous nights; to guard the San Juanico Strait in case the Japanese navy decided to sneak through. But little did the men know that the Japanese had no intention of entering Leyte Gulf through San Juanico Strait at the northern end of the gulf.

That's because major remnants of the Japanese fleet instead intended to enter the gulf through Surigao Strait at the southern end of Leyte Gulf. And they were already on their way, steaming full speed ahead. The Allies were only hours away from engaging with the Japanese fleets in what would become the largest naval battle in history. The Battle of Surigao Strait was about to begin.

Truth be told, no previous amphibious operation to date in the Pacific Theater had brought together so many tempting targets for the Japanese as there had been gathered in Leyte Gulf starting on A-Day. The sheer amount of bait for the Japanese that sat in Leyte Gulf by the end of A+1 Day was stunning: 151 LSTs, 79 LCIs, 58 transports, 221 LCTs, and hundreds of other vessels steamed in and out or lay anchored in the Gulf.[35]

The massive Japanese fleet that had assembled on its way to the Philippines was fully aware of the feast that awaited them if they could fight their way into Leyte Gulf. The entire Japanese fleet approaching the Philippines was divided into four major groups. In the days leading up to October 24, it became clear that the main body of the Japanese fleet was headed for a two-pronged attack on the shipping in Leyte Gulf.

One prong, the Southern Force, was headed for Surigao Strait— the southern entrance to Leyte Gulf. The other prong, the Center Force, was headed for San Bernardino Strait. The Center Force intended to pass through it and swing to the southeast and entering

Leyte Gulf through the same channel that the Allies had used on A-Day. Allied planes and submarines had been pummeling both the Southern and Center Forces, but the Japanese kept on coming.

Shortly after midnight on the night of October 24–25, the flagship *Blue Ridge* received word that one of those groups of enemy ships had entered Surigao Strait and were headed north. This group of Japanese ships heading toward them was Admiral Shōji Nishimura's Number One Diversion Task Force. Nishimura's force was one of two groups that made up the Japanese Southern Force. Nishimura and his men wanted to crush the American ships, which included Stephen's LCI.

"[T]his particular one here, their Southern Fleet," Stephen said as he pointed to the Surigao Strait on the map in front of him, "the Japs came up and . . . in this position was this Oldendorf with all our battleships, and when the Jap fleet walked in, [Oldendorf's ships] opened up."

Luckily for Stephen, the Allies were prepared to meet Nishimura's fleet at the mouth of the Surigao Strait. Earlier in the day of October 24, Rear Admiral Jesse B. Oldendorf, aboard his flagship USS *Louisville*, had moved the Allied ships to the southern end of Leyte Gulf at the mouth of the Surigao Strait. There, they began a steady patrol back and forth across the channel, laying a trap for any Japanese that might be foolish enough to try and enter Leyte Gulf from the south. Admiral Oldendorf not only had superior firepower, but his heavy ships were in a perfect tactical position. From where they were positioned, Oldendorf's line of battleships, cruisers, destroyers, and PT boats could cross in front of the line of approaching Japanese ships, allowing Oldendorf's giant ships to aim all their guns directly at the enemy ships, while only receiving fire from the forward guns of the Japanese ships. This is a classic naval warfare tactic called "crossing the T" that every Navy high commander dream of pulling off.

And sprang his trap he did: in a masterful and brilliant fashion, Oldendorf executed his plan to perfection. In the early morning hours of October 25, Oldendorf and the Allied battleships, cruisers, destroyers, and PT boats successfully "crossed the T" annihilating the

Japanese column of ships as they approached. With every single American ship firing at the oncoming Japanese with every single available gun, the battle reached a spectacular climax from 3:50 to 4:20 a.m. in which Nishimura's armada of ships was absolutely decimated as they approached Leyte Gulf through Surigao Strait. This included Nishimura aboard his flagship *Yamashiro*, who also perished in the battle when the *Yamashiro* sank from an explosion caused by two torpedo hits.

And yet, the Japanese continued to charge up the strait into the trap prepared for them. A short time after that, the second group from the Japanese Southern Force, the Number Two Diversion Task Force led by Admiral Kiyohide Shima, followed the same path of Nishimura's fleet up the Surigao Strait, and was met with the same fate by the same Allied ships.

"Admiral Oldendorf had six battleships, he was the main guy, . . . and that was a hell of a battle," Stephen remembered.[36] "They had that formation they used to say, 'Cross the T'—in other words, you get in range of these battleships," he said while pointing to the battle map and imitating a time-out motion with his hands, "and they had all the guns on [the Japanese] . . . and Oldendorf is the guy that should get the credit, 'cause he crossed the T. And anything that comes in, all their firepower [is] right directly on the Jap force. . . . They had all the firepower of every ship on them. That's the reason they got them." Stephen then added that off in the distant southern horizon, "We watched it all night long when they were fighting."

By 5:05 a.m., while the LCI 65 went to General Quarters for their first condition red of the morning of October 25, (A+5 Day) the straggling remnants of the Japanese Southern Fleet turned around and began retreating south down Surigao Strait. Stephen remembered seeing the battle unfold saying, "The next day, [the Allies] knocked them out here with their fleet—our fleet . . . I watched that sea battle." He laughed as he described how he felt seeing American battleships shell the enemy, "That's all you see is '*va-doo! va-doo! va-doo!*' And I thought, man this is fun!" Stephen had no idea that he was witnessing the true end of an era in naval warfare. With his own two eyes, he watched in awe the world's last engagement of a battle line between two naval fleets in which air power played no part.[37]

With their chance to completely finish off the entire Japanese southern naval fleet, Admiral Oldendorf and the American ships began chasing after what remained of Admiral Shima's fleet down Surigao Strait. Stephen and the rest of the crew cheered them on with rowdiness. Stephen remembered a shipmate said to him, "'I feel sorry for the Jap survivors, 'cause if they swim to the shore in the Philippines, [the Filipinos] were waiting for them.'" The Filipino guerillas were known for how much they despised the Japanese.

However, at that moment, Stephen also had no idea that this move by Oldendorf now left the entire amphibious landing force in Leyte Gulf—himself included—completely unprotected and in mortal danger all thanks to Admiral Halsey.

———

The single most controversial decision in the Battle of Leyte Gulf had been Admiral Halsey's decision to unexpectedly abandon the amphibious landing force that he was supposed to be protecting in Leyte Gulf. He had taken the more than ninety warships that made up the entire weight of the Third Fleet carrier force to engage a third Japanese fleet—Admiral Jisaburō Ozawa's Northern Task Force.

At about the time Stephen's LCI had been laying a smoke screen in Leyte Gulf the night before, Admiral Halsey had received reports aboard his flagship USS *New Jersey* that the enemy's Northern Force had suddenly been sighted. With his disdain for cautiously waiting, Halsey took advantage of a split command and decided—impulsively and predictably—to attack this Northern Force near Luzon in what became the Battle off Cape Engaño. But Halsey was unaware that his decision was playing right into the hands of the Japanese, who had set a trap of their own. It had been Ozawa's intention all along to lure away all of Halsey's battleships and cruisers from San Bernardino Strait, and Ozawa knew Halsey would take the bait.

"It was in fact little more than a decoy, deployed specifically to lure Halsey out of position," noted renowned author and historian James D. Hornfischer. Halsey abandoned his post, according to Hornfischer, "in favor of attacking a squadron of carriers that, with a sparse complement of planes, presented little threat to the U.S. beachhead."[38] In the end, Ozawa's Northern Force turned out to be only nineteen Japanese ships.[39]

"Halsey committed a fundamental error in forming his conclusion. He determined ahead of time what he wanted to do, then fit events into a prearranged pattern to conform to his wishes," author John Wukovits wrote of Halsey's blunder.[40]

The purpose of Ozawa's trap to lure Halsey away was to clear the way for a fourth Japanese fleet group—the Central Task Force under the command of Admiral Takeo Kurita—by far the mightiest and most dangerous of the Japanese fleets in terms of firepower. Shortly after midnight, Kurita's Central Force steamed out of the San Bernardino Strait unopposed, unharmed, and undetected toward their target: the U.S. Seventh Fleet.

For the next few hours, Barbey and his men watched from aboard the *Blue Ridge* as occasional flashes of gunfire broke through the darkness. The air was full of dispatches and all the news was good news during the battle of Surigao Strait until the Allies began receiving other dispatches telling a different story. The San Bernardino exit was no longer being guarded. Halsey had withdrawn all his ships and was chasing six Japanese carriers, three cruisers, and some destroyers that had come down from the north.

Based off his aviators' reports, Halsey had mistakenly believed that the Japanese Center Force had been so battered in the Battle of the Sibuyan Sea from the previous day, that they would not attempt passage through the Strait. But he was wrong. The Allies learned later that this Northern Force had been sent down to decoy Halsey away from San Bernardino. Halsey caught the decoy and crushed Ozawa's Northern Force. The Japanese plan had worked.

———

"If they'd ever got in there, we would've been dead ducks," Stephen recalled of Kurita's Central Force that exited San Bernardino Strait and headed for Leyte Gulf, which kicked off the Battle off Samar. He pointed to San Bernardino Strait on the map in front of him and continued, "We were lucky because like over here their Central Fleet came down, and man, I look out like this," he said as he held his hands up to his eyes imitating binoculars, "and you could see the Japs." He laughed and added "Their destroyers out there, I thought aw man is this for real?" When asked if his LCI in Leyte Gulf could've fought back against the heavy ships of the Japanese Center Force, Stephen replied, "No, that was too far out for us. That was the big ships that handled that. Ours was a little bitty boat. Hell, we couldn't even make a dent in one of them."

At around 7 a.m., the LCI 65 went to General Quarters in what had become an extremely critical situation for the Seventh Fleet. Japanese ships were now attacking the American escort carriers just outside of Leyte Gulf guarding the main entrance to Stephen's east. The Japanese were seriously threatening the entire amphibious operation at Leyte. The Allies sent emergency dispatches to Admiral Halsey calling for him to turn around some of his fast battleships to help.

As the dispatches shot off to Halsey, a little further up the San Juanico Strait nearby, fellow LCI Group Forty-Five gunboat, LCI (G) 23, suffered massive damage from a Japanese bombing attack.[41] A group of three Japanese "Tony" planes and one "Val" plane were spotted at around 7:55 a.m., and two of them crashed near enough the LCI 23 to kill several men and cause major damage to the ship. The "Val" attacked the LCI 64 (which was present with the LCI 23) and scored a near hit on the LCI 64's stern. The LCI (R) 338 came to the LCI 23's rescue, as it delivered fire fighters and pharmacist mates to help the LCI 23 just in time. By 10:45 a.m., with all the wreckage on the LCI 23 cleaned up, a salvage party was sent over to help make her seaworthy again.[42] Motor Machinist's Mate 3/c Charles Ports remembered being injured in this attack saying, "I was wounded in both knees, both thighs, both buttocks, groin, left forearm and wrist."[43]

Nearby aboard the *Blue Ridge*, Captain Tarbuck made the following notations in his official journal:

0900: People here feel that Halsey's Third Fleet battleships are chasing a secondary force, leaving us at the mercy—of which there is none—of the enemy's main body.

His report urgently and tensely went on:

If our analysis is faulty it is because we are the ones who are trapped in Leyte Gulf. As soon as the Jap finishes off our defenseless [escort carriers] we're next, and I mean today.[44]

Meanwhile, a different battle was taking place near the eastern entrance of Leyte Gulf, just off Samar Island, as Japanese battleships, cruisers, and destroyers began to clash with Rear Admiral Thomas L. Sprague's "baby flattop" aircraft carriers from among his carrier task units.

At around 8:50 a.m. on October 25, exactly twenty-four hours after Aiston had been killed, the LCI 65 witnessed the Allied ships in the bay begin to fire upon three Japanese bombers they'd spotted. The daily Japanese air attacks had begun again. The bay erupted in gunfire. One Japanese Betty bomber was brought down, but it released a bomb that hit the LST 552 on its way down, seriously damaging the LST's port side and killing five men.[45] From that moment on, for the remainder of the day, the 65 spent its time constantly under air attack fighting Japanese dive bombers.

But Stephen and the rest of the transports and smaller ships of the Amphibious Force in Leyte Gulf would end up getting very lucky. To their east, Rear Admiral Clifton "Ziggy" Sprague's heroic but hopelessly outnumbered ships of task unit "Taffy 3" fiercely fought back against the Japanese ships of Kurita's Central Force that tried to enter Leyte Gulf, and won the day.[46] In his history, Morison wrote of the Battle off Samar: "In no engagement of its entire history has the United States Navy shown more gallantry, guts and gumption than in those two morning hours between 0730 and 0930 off Samar."[47] It was truly one of the most remarkable and epic battles in naval warfare. Astonishingly, Kurita's mighty squadron was forced to turn and flee because of the heroics of Sprague's small task unit.

In a fatal error, Japanese Admiral Kurita decided to turn around and retreat back into the San Bernardino Strait from which his Center Force had come, instead of smashing the exposed and defenseless Seventh 'Phibs in Leyte Gulf. Even after pummeling Taffy 3, Kurita was spooked when he learned what had happened to Japan's Southern Force by Oldendorf's fleet the night before in Surigao Strait. Kurita had also encountered an aggressive fight from the Americans off Samar—causing him to overestimate their forces— which only reinforced his belief that retreating was the only option. The heroism of the American boys who fought in the Battle off Samar is the reason why every single amphibious ship in Leyte Gulf —including Stephen's LCI 65—escaped what would have surely been a slaughter if Kurita's Center Force entered the harbor. Stephen and his LCI had been saved by Taffy 3.

LCIs later became the unsung heroes to the American survivors of the ships that had bravely charged headfirst into the thick of the main Japanese fleet and been sunk.[48] More than eleven hundred sailors floating in the shark-infested waters off Samar—like those from the destroyer escort USS *Samuel B. Roberts*—were rescued by the small LCIs over the coming days. In the words of one grateful survivor from the *Roberts* who was rescued after spending two days stranded at sea, "Some people make fun of the LCI and say she's clumsy-looking and ugly," began Chief Yeoman Gene Wallace, "but to us on the raft and net this was the most beautiful sight we had ever seen."[49]

The crew aboard the LCI 65 would spend the next four days, through October 29, under constant Japanese air attack and a non-stop cycle of General Quarter red alerts. It can be safely assumed that none of the crew was able to get sleep due to the constant running back and forth from their bunks to their battle stations at all hours. Nearby on the LCI 440, Dominick Maurone remembered, "We remained in the bay between Leyte and Samar as the Japanese air force went all out with kamikazes. They came in day after day, and from our battle stations we watched as they dove at our big ships, hitting many of them."[50] Russell Hartwell, who also served in the Philippines aboard the LCI 711, once wrote of the constant kamikaze threat: "At times, as fast as you could look around at different places in the convoy, you could see Jap suicide planes crashing into American ships or into the sea. Many men died and too many ships went to the bottom or were destroyed. . . . We gave up on the idea of hitting the sack after awhile as the attacks came too often and too fast. We tried going to bed at first, but as soon as we would get into our bunks, general quarter's alarm would sound and we would have to rush back to our battle stations."[51]

———

On the evening of October 29, the crew of the 65 received news of a brand-new threat. This one had nothing to do with the enemy. In his war diary for LCI Group Forty-Five, Commander Archie "Squad Don" Holmes wrote:

> On the night of 29 October a Typhoon reached LEYTE-SAMAR area during which the LCI(G) 23 drifted and finally broached on [CATAISAN] POINT, just south of the newly constructed airstrip. The LCI(G) 65 had her anchor cable cut by a vessel that drifted on to her and, later broached on RED BEACH, LEYTE.[52]

The LCI 65 had been assisting and towing the LCI 23 that evening, but quickly learned from reports that a typhoon was coming. In a hurry they cast off their lines from the LCI 23 and proceeded to anchor right off Tacloban airstrip near White Beach and took all precautions against the foul weather. But the intensity of the typhoon began to increase considerably at around 1:20 a.m. Ten minutes later, the crew of the 65 noticed their anchor cable was dragging the sea-bottom, so they started their main engines in an attempt to hold their position to prevent the anchor cable from severing. Five minutes later, caught up in the chaotic scene, a nearby unknown LCI (L) collided with the bow of the LCI 65 causing a large hole to penetrate through the hull of the ship and forward void tank. As the 65 swung around to try and avoid the rogue LCI, the 65 banged against the side of the unidentified LCI several more times which caused the 65's forward anchor cable to snap. There was also minor damage inflicted on the starboard side of the 65. Five minutes after that debacle, it became apparent that the 65 would need to flee for the beach, which would be the safest place for the small LCI gunboat since she could no longer anchor.

By 2 a.m., the 65 made it to White Beach off the eastern coast of Leyte, but the heavy swells blew the LCI onto the beach. The swells smashed the port side of the 65, punishing her for several hours as she lay broached (stranded) on land. It wasn't until about 6 a.m. that the intensity of the typhoon winds started to die down a bit.

That typhoon was the event that Seaman 1/c J.R. Reid remembered most from the Leyte campaign. "We had a couple of typhoons [that] blew in on us while we [were] there," he said. "We got broadsided on the beach and it took 'em about five or six days to get us off."

At 8 a.m. on October 30, the LCI (R) 338 attempted to cast lines and take the 65 in tow, but they were unsuccessful and unable to come close enough. About an hour later, the crew of the 65 struggled to move about because the port side of their gunboat was tilted about thirty degrees and was buried in five feet of sand, with their starboard side sticking up in the air. By 10 a.m., Macaluso prepared to have his crew temporarily abandon ship until the 65 could get towed off the beach. About an hour later, an LCM with towing gear from the tugboat USS *Quapaw* commenced transferring towing gear onto the 65. Shortly after noon, the crew began to fasten and secure the towing gear to the forward parts of the 65. Around 7:45 p.m. another ship, USS *Chickasaw*, also began fastening and securing towing gear to their ship in an effort to help. It wasn't until after sunset that the towing gear was completely secured, and the first attempts were made to pull the 65 off the beach. By 11:45 p.m., the

Chickasaw and LCMs from the *Quapaw* only managed to pull the 65 ten feet.

After spending all night broached on White Beach, the crew of the LCI 65 awoke to the blaring whistle of General Quarters at 5:45 a.m. on October 31. A lone Japanese bomber approached the stranded 65 and dropped a bomb near them but missed.[53] After some discussions with an officer from another ship, the 65 tried something else. At 7:36 a.m., the 65 began pumping their fresh water over the side of the ship, and an hour later the crew also transferred much of the 40mm ammunition to the fantail of the ship, shifting the weight of the ship to make it easier to pull off the beach. As they were doing this, the LCI (R) 341 pulled up alongside the 65 and began spraying the 65 with a high-pressure water hose in an effort to wash away as much sand on the port side of the ship as they could. After a couple hours, several more vessels came alongside to help spray the sand off the side of the 65 buried in sand. By the early evening, the towing operation was temporarily put on hold as the crew of the 65 rushed to their battle stations in response to a Japanese air raid. Several hours after the air raid passed, the official records reflected that the 65 had moved seven degrees seaward. However, as the *Chicksaw* attempted to pull the 65 off the beach, the tow line snapped.

So there, in the closing days of October, stranded and helpless on Leyte Island's White Beach, Stephen and the crew of the LCI 65 sat broached waiting for a new tow line to be secured.

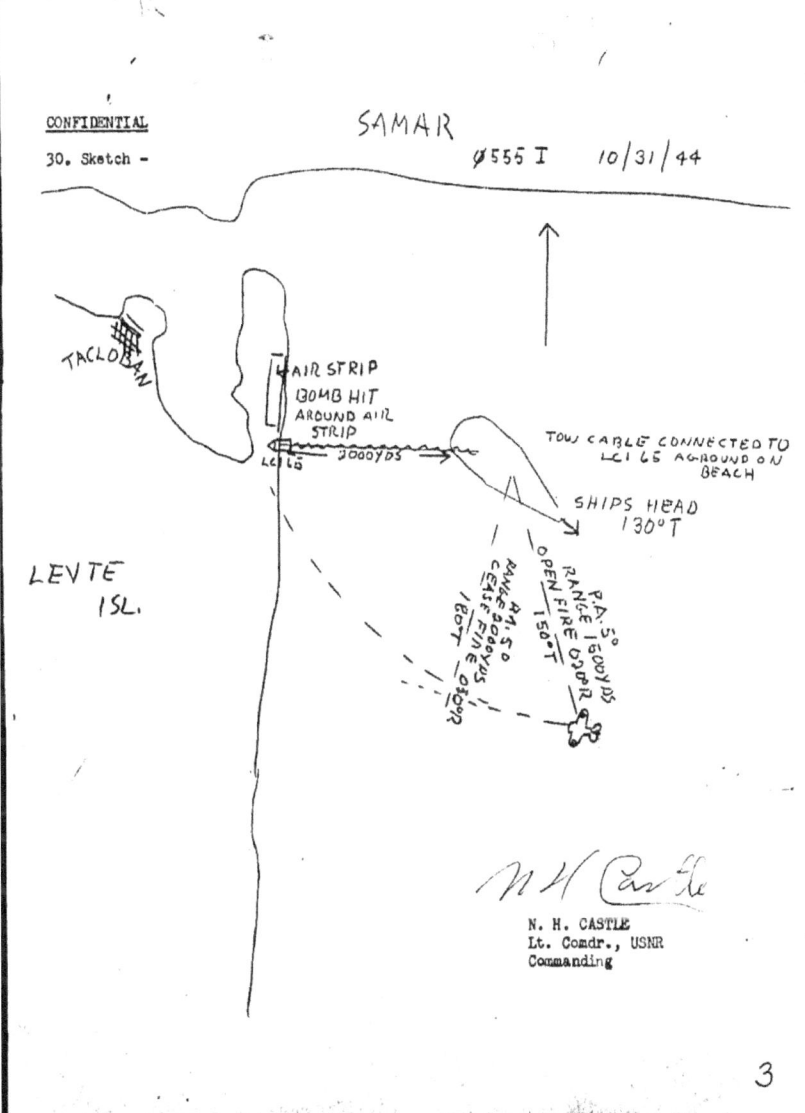

A sketch from the USS *Quapaw* detailing attempts to pull the
LCI (G) 65 off White Beach, Leyte Island, October 30-31, 1944.
(U.S.S. *Quapaw, Action Reports – Philippines Operation – For
Period 29 October to 14 November 1944*, p. 5)

17

NOVEMBER 1944: BLOODY SUNDAY

"I think that the expectation and waiting for the attacks were as hard on our nerves, if that's possible, [as] the actual combat."

– Russell W. Hartwell, LCI 711 [1]

Stephen and the 65 remained stranded high and dry on White Beach to begin the month of November as the ships around them worked to install another tow line to the gunboat. With the winds of the typhoon significantly died down by the morning, the *Quapaw* had another tow line secured to the 65 by 7:45 a.m. Unfortunately, the second cable also snapped, and a third attempt was begun to secure a tow line. In the middle of the process, the men hurried to their battle stations as the General Quarters alarm whistled a condition red. As the men stood at their posts, an LCM came alongside to try their luck on securing a tow line, which would take them the rest of the day. Some crewmen were permitted to sleep on the beach.

At 6:15 the next morning, the crew members who had slept on the beach reported back aboard the 65 and resumed regular duties. The crew went back and forth to their battle stations as General Quarters was sounded several times that morning. Shortly before noon, the *Quapaw* once again attached towing gear to the 65's bow. Later in the day, verbal orders were given from the Army Medical Officer to transfer Motor Machinist's Mate 3/c Gerald L. Edwards to the medical ship LST 464 to treat his burns that he suffered from a gasoline fire on the beach. After several hours passed, the LCM and the *Quapaw* secured their end of the tow line, and for a third time, the tugboat attempted to pull the 65 off White Beach. The 65 began to slowly move! Inch by inch, beginning at 7:19 p.m., the tugboat successfully pulled the 65 off the beach. By 7:44 p.m., the ship was clear of the beach and afloat in Leyte Gulf once again. After four days of being stranded on the beach, and three different tow lines later, the little gunboat fired up her engines and began moving on her own toward the rear of the *Quapaw*, still in tow.

The moment the clocks struck midnight local time on November 3, the crew of the 65 had been at their battle stations for nearly four hours. Twenty minutes after midnight, the crew was secured from General Quarters, and the men returned to their bunks, but as the men began to doze off, the General Quarters whistle blared, waking them suddenly an hour later. The men scurried as quickly as they could in the darkness, up the steep ladder that led topside, and to their battle stations, but within fifteen minutes, the condition green "all clear" was given. The weary men returned to their bunks. Not even a full half hour later, the General Quarters whistle pierced the black air, with the crew once again returning to their battle stations. They remained there for nearly an hour and a half before the "all clear" was given again. The exhausted, haggard men headed back to their quarters and climbed back down the steep ladder, one by one, and laid down in their bunks. If any of the men were lucky enough to catch some shut eye by that point, the General Quarters alarm that blared yet again at 4:30 a.m.—for the fourth time that morning— ensured that wasn't going to happen any longer. This lifestyle had become the norm for every young man the moment they had entered Leyte Gulf. The worn crew remained at their battle stations for about an hour before the first Japanese planes of the day were spotted. At 5:30 a.m., the ships in the harbor opened up on the enemy planes signaling the start of a new day. An hour and a half later, the crew was given the condition green to return to their duties. Shortly after, the LCI 69 came alongside the 65's port side to take them in tow for a short period of time, in order to relieve the *Quapaw*. After casting off all lines, the 65 got underway to a different spot. At 10:40 a.m., the LCI 70 came alongside the 65 to bring aboard supplies that the 65 was running low on. After all supplies and stores were brought aboard the 65, the LCI 70 cast off all lines and got underway. About an hour later, a crewmember who had slept on White Beach and failed to reboard the 65 the previous day when they were pulled off, Seaman 1/c Frank Hook Jr., returned to the ship. It is unknown why Hook had not returned to the 65 by the time they were towed off the beach, but by 1:55 p.m. he had turned up again safe and sound. Just after sunset, the 65 got underway to change its position, eventually anchoring in San Pedro Bay. And in the same manner they had begun November 3, so is the way they ended it, with the crew going to General Quarters four different times before the clock struck midnight.

November 4 began with the men at their battle stations in the waters off tiny Dio Island across from White Beach. The men remained there almost all through the night. The sleep deprivation was starting to take its toll on the men, as they were growing irritable and irascible. At around 4:30 in the afternoon, a captain's mast was held for new crewmember Gunner's Mate 2/c James O. Vincent where he was sentenced to five days of solitary confinement for insubordination. Vincent was one of the four men who had joined the 65's crew back in New Guinea right before they departed for Leyte. A couple hours later, the records and gear for Motor Machinist's Mate 3/c Gerald L. Edwards—who had been sent to a medical LST two days earlier to treat his burns—were also transferred to the LST 464.

November 5 and 6 were routine days in the waters of San Pedro Bay, filled with air raids and General Quarters whistles that blared the men to their battle stations numerous times throughout the day. They also managed to resupply freshwater courtesy of the LCI 68.

The next day the LCI 69 came alongside the 65 to pick up their outgoing mail, as Stephen and the crew spent yet another day in San Pedro Bay going back and forth to their battle stations with each sighting of Japanese planes. On Leyte Island, the foot soldiers of U.S. Sixth Army's X Corps were driving down the Ormoc Valley and were engaged in some of the toughest and bitterest fighting in the Leyte campaign, according to Morison's history.

On November 8, the men awoke to their first condition red of the morning at 5 a.m. but were given the "all clear" about a half hour later. However, very shortly after that, Stephen and the crew were given news from the States that President Franklin Roosevelt had defeated Republican Thomas E. Dewey and won an unprecedented fourth term in yet another landslide election. Besides that, and some issues with their ship's anchor dragging, the crew had a routine day in San Pedro Bay. On the afternoon of November 9, Gunner's Mate 2/c James O. Vincent was released from solitary confinement after completing his five-day sentence. November 10 and 11 were routine days in San Pedro Bay, going back and forth to their battle stations eight different times during that twenty-four-hour period, but no ammo was fired.

The crew was relieved to learn that the next day, a familiar friend would be repairing a hole in their ship. The job was assigned to landing craft repair ship USS *Achilles,* which was actually just the re-outfitted LST 455—the same LST that had conducted repairs on the 65 in Alexishafen back on the night of August 19. According to J.R. Reid, "We got a hole [from] the typhoon, I guess it was a cargo ship [that] broke loose and they knocked a hole in [LCI 65] . . . and that's the reason we [were] over there getting that hole patched up."

It would be a day the crew of the 65 would never forget.

———

The USS *Achilles* (ARL-41) was originally commissioned as USS LST 455, but because of the immediate need for additional repair ships in the Pacific, the LST was converted into an Achelous-class landing craft repair ship. The LST had its entire interior redone so that it could be filled with machines used for repairing all kinds of Allied wartime instruments. It was one of several former-LSTs-turned-repair-ship that were present in Leyte Gulf on the morning of Sunday, November 12, 1944, a morning where the LSTs were busier than usual—much busier. The LSTs consisting of variations of repair ships, liberty ships, and medical ships filled Leyte Gulf that morning. Sea traffic crowded in all directions. Many of the LST repair ships, which included the *Achilles*, were attending to smaller LCIs that had been damaged from the typhoon that struck on October 30—one of which was Stephen's gunboat. According to *Achilles*' official action report:

> On the afternoon of 12 November 1944 we were anchored in berth 55 San Pedro Bay, Leyte, P.I. with LCI 1056 along our port side and LCI 432 along our starboard side forward, LCI 65 starboard side aft.[2]

———

"It was a repair ship," J.R. Reid remembered about the *Achilles*. "I think there were four LCIs tied up, two on each side," he recalled of the arrangement that morning.

The LCI 65 pulled alongside the starboard side of the USS *Achilles* at 8:35 a.m. on the morning of November 12. Stephen's gunboat was also alongside the rocket ship LCI (R) 338, which was right next to the repair ship, but was scheduled to depart shortly so that the LCI 65 could take its place. The LCIs 432 and 338 sat comfortably ahead of the 65 also on the starboard side of the *Achilles*. Stephen remembered of that morning, "We [were] tied up alongside this LST, and there was another LCI in front of [*Achilles*], and [LCI 338] was getting underway, and we [were] gonna pull up." While the 65 waited for the LCI 338 to depart, a repair officer from the *Achilles* came aboard the 65 to assess the damage. In less than ten minutes, the repair officer departed the gunboat. Macaluso left the ship at 10:05 a.m. but quickly returned fifteen minutes later.

Then, at 10:55 a.m., the LCI 338 finally cast off all lines and got underway. At exactly 11:00 a.m., the 65 pulled forward and took its place on the starboard side of the *Achilles*, right behind the LCI 432. At noon, there was a daily inspection of the ship's ammunition magazines, all of which were reported to be in normal, working condition. About twenty minutes later, the LCI (L) 28 joined the 65, where it moored alongside the starboard side of the 65 but ten minutes later cast off its lines and quickly departed, because a Japanese air raid was beginning in the distance. At 1:30 p.m. the General Quarters whistle aboard the 65 blared its battle cry sending the crew scrambling to their battle stations. Stephen joined his skipper Macaluso atop the conning tower.

USS LST 455 on March 7, 1943. It was eventually converted to a landing craft repair ship, being renamed USS *Achilles*. The ship's casualties totaled 19 killed, 28 wounded, and 18 missing on November 12, 1944. (U.S. National Archive photo No. 19-N-42328)

According to the official war diaries and deck logs of the various ships present that day, three Japanese Zeke fighter planes suddenly appeared from the clouds at 2:20 p.m. and approached the *Achilles* bearing 010 degrees relative. The Japanese fighters were already extremely close by when they dipped below the cloud line, so the Allied ships in San Pedro Bay were caught by surprise and didn't have much time to react. Stephen, however, with his self-described sharper than usual hearing abilities, had heard the planes coming before they appeared. His ears heard a rumble in the distance that he recognized immediately. He recalled commenting to Macaluso, "I told the old man, I said, 'man, there's some Jap planes up there.'" Macaluso dismissed his warning, but Stephen insisted, "'There's a Jap plane up there!'" He added, "No sense in opening up my mouth, and here he comes."

The men aboard the *Achilles* suddenly spotted three Japanese Zekes that quickly appeared out of the low-lying clouds to their north. The center plane broke off from formation and went into a hell dive headed straight for the *Achilles*, firing his guns furiously as he dove.

Lieutenant M. D. Coppersmith, an officer aboard the LCI 432, wrote in his official action report detailing the beginning of the attack:

> USS LCI (L) 432 was moored alongside starboard side forward U.S.S. Achilles, for availability for accomplishment of repairs, in San Pedro Bay, Philippines, on the 12th day of November, 1944. At approximately 1400 three planes came out of a cloud dead ahead, elevation approximately 050° and went into a suicide dive.[3]

Boatswain's Mate 1/c Benjamin Schmidt, the gunner on the LCI 432's number one 20mm gun, completely shot off the tail of the Zero when it flew within two or three hundred yards, just as the gunners on the *Achilles* began to fire. The gunners aboard the LCI 65 quickly opened up with their own three-inch 50 caliber, 40mm, and 20mm guns, firing furiously at the three Japanese fighters. Lieutenant Coppersmith's report went on:

> The center plane came directly forward of [t]his vessel, passed between the conning tower of this vessel and the forward boom of

the USS Achilles and crashed into the forward portion of the deck house of the USS Achilles.[4]

The other two Japanese pilots broke off and picked their targets as well. The LCI 432's report noted:

The Zero which came in on the right side of the three, passed to our starboard under fire of our number five 20mm gun and was later observed crashing into a Liberty ship about one thousand yards astern of our vessel.[5]

At 2:21 p.m., the Japanese pilot of the center plane cemented his place forever in the historical lore of World War II as a kamikaze. Once he began his dive, it only took fifteen seconds for the enemy pilot armed with a small anti-personnel bomb to crash into the forward deckhouse of the *Achilles*. It would be one of many kamikazes that would rain from the sky that horrific afternoon in Leyte Gulf. The pilot crash dove into the middle of the repair ship, hitting it right next to where the LCI 65 was tied up.

"Man, that was no more than fifty feet from me where the plane hit!" Stephen recalled.

As soon as the kamikaze struck the *Achilles*, a massive explosion followed, consuming the ship's carpenter shop, enveloping it in an inferno. A chaotic scene ensued. Stephen remembered when the kamikaze hit the *Achilles* saying, "He smacked right into that LST. And you could see all them kids up there getting nailed."

J.R. Reid sadly recalled that the *Achilles* had a massively exposed floor deck in the middle of the ship that "had a big opening in the center . . . and they had a shop set up down there, and this [kamikaze] went down in there—in that hole—blew that thing out." Reid added, "He didn't sink it, but it did mess it up pretty bad, and there [were] quite a few guys that got killed on it." Since the *Achilles*' crew had been inundated with ships to repair that day, many of the men had been reluctant to evacuate the floor deck, because they didn't want their work interrupted.

According to the *Achilles'* official action report, all hands performed in a commendable manner fighting the fires that broke out on their ship, but it was a General Motors technician named Ray Dunwoody who stood out the most that Sunday afternoon. Showing complete disregard for his own life, Dunwoody grabbed a fire hose, and fearlessly sprinted up to the top of the deckhouse—amid exploding 20mm ammunition—to rescue men from the flames. He somehow managed to remove several seriously wounded men all while keeping his hose in action against the fire the entire time. The official report from the *Achilles* detailed Dunwoody's bravery by stating:

> His actions and leadership were an inspiration to everyone. It is recommended that his action receive special recognition.

The report from the *Achilles* went on to commend one of her brave shipfitters wounded from the fire:

> In doing this, [Dunwoody] was assisted by Wayne H. BASS, SF3c., although badly burned about the hands.

Alongside the *Achilles*, the LCI 432 immediately joined in the fight against the fires and assisted with the wounded. By 3:00 p.m., LCIs 72, 73, and 338 had come alongside with fire hoses to also lend assistance. Men of valor aboard the LCI 432 displayed a heroic effort fighting the fires on the *Achilles*, even as fires on their own LCI broke out. According to the LCI 432's action report:

> A sizable fire broke out on the Achilles. This vessel remained along-side to render assistance in fighting the fire and to aid casualties. When the fire main pump became operative, one stream of water from our fire hose was played on the fire forward on the weather deck of the Achilles. A few minutes later, two handy billies from this ship were pumping streams of water to the same fire. A small fire which broke out among the smoke pot stowed on our forward well

deck was extinguished with CO_2 extinguishers. Other small local fires were extinguished by hand.[6]

In stark contrast to the assistance lent by other LCIs, however, Stephen's LCI did not help the burning, ailing *Achilles*. According to Stephen's account of that day, this was because his captain, Macaluso, allegedly refused. Stephen later recalled this incident from the 65's conning tower explaining that his LCI had been right alongside the *Achilles*, and when the kamikaze crashed into it, he could see a bunch of the men from the *Achilles* struggling in the water. Stephen said he turned to his skipper and asked, "You want some of our guys to go over there and give them a hand?" Stephen was shocked by the response, "[Macaluso] says, 'No, we're getting the hell out of here!'" Stephen painfully recalled this moment almost sixty-seven years later, as he never forgot about his friends from the *Achilles* that he witnessed perish in the water. As Stephen's LCI fled the scene, he wished he could have helped save the men, especially the ones burning, but all he could do was watch them get smaller and smaller in the distance until they disappeared from view.

One of those boys was John Angle. Angle had been one of the men who helped repair the LCI 65 in the middle of the night nearly three months earlier at Alexishafen. He had joined the Navy on November 12, 1943. In a cruel and tragic twist of irony, John Angle died on his one-year anniversary of enlisting. At 2:24 p.m., the official entry in the deck log of the LCI 65 by the engineering officer, Ensign Elmer Kinsinger, simply stated:

Unable to render assistance. Got underway, various courses and speeds.[7]

"He was a chicken shit son of a bitch," Stephen bitterly recalled of Macaluso's decision to retreat and abandon the young men from the *Achilles* trying to survive in the water. As the skipper, Macaluso decided to get his men out of there as quickly as possible, while the LCI 432 and others helped extinguish the raging fires still blazing aboard the *Achilles*. "I felt so sorry for them kids," Stephen sadly reflected. "Blew 'em all apart. They [were] just killed on the deck. There might've been parts . . . that blew on our ship from them kids," he said as he looked down with an agonizing sadness in his voice. Twenty-four men from the *Achilles* were killed that day, including three passengers from LCT Group Nineteen and Twenty-One staff, and thirty-five men were wounded.[8] Being at the 20mm gun at the back of the ship, J.R. Reid did not remember seeing the men from the *Achilles* that Stephen saw. Reid did however remember, "Everybody was so scared they didn't want to do anything then. Cause it was—it was pretty wild." The only additional notes offered by Macaluso himself from that day come at the very end of his anti-aircraft action report and were very brief:

> This is the second time this vessel has undergone a suicidal crash.[9]

At that point in the afternoon, the sky had only just begun to rain kamikazes in all directions. It looked like some twisted version of an aerial banzai charge. Nearby, LCI (L) 684 was also hit by a kamikaze and effectively sunk.

Lieutenant R. Y. Martin, commanding officer of arguably the most important medical ship in the harbor, LST 464—the ship that had disposed of Aiston's body when he was killed on October 24—later wrote in his official action report detailing the chaos of November 12, 1944:

> . . . enemy aircraft began suicidal diving on shipping; two crashed off bow of this ship; three crash-dived Liberty ships; and the rest missed their targets; this suicide attack lasted about five minutes, but they were the wildest and most horrifying minutes ever spent by this command . . . [10]

The medical officer of LST 470, another LST present in Leyte Gulf, recalled a memorable story from that afternoon:

On bloody Sunday, 12 November 1944, while on the Leyte supply run, we had our first experience with Kamikazes. When the shooting started I went back to the liquor locker and selected a good quarter of bourbon that I had hoarded, and reappeared on the deck. I did the best I could to give everyone a "nip." That's all they got, but it was a whale of a morale builder. . . . The Executive Officer told me afterwards, "Doc, I was green until you came along with the bottle of booze. I was a man after one small snort. I could kill them all."[11]

The staggering death toll and damage that resulted from the kamikaze attacks throughout the day on Allied ships cannot be accurately conveyed in this book. Among that carnage, six other freighters were hit by kamikazes, but they ultimately returned to the west coast of the U.S. under their own power.[12]

Lieutenant Martin also ended his action report with the stirring words of his crewmen's courage. He concluded his official action report on the events during the Philippines campaign by stating:

All hands worked tirelessly in protecting and ministering the patients. When ammunition became low at the guns, the hospital corpsmen passed ammunition; when patients were being received or transferred, deck hands and engineers and all hands carried stretchers and helped make the patients comfortable; when sick bay overflowed which it did frequently, all hands gave up their bunks for the steel decks topside so the patients might be cared for.[13]

For the rest of Bloody Sunday, Stephen and his shipmates aboard the LCI 65 would remain at their battle stations long into the night, haunted by the scene they had just witnessed.

———

Meanwhile, back in Hollandia, Admiral Barbey and the rest of the Allied high command were busy planning for their next targets in the Pacific War. Although it would take the combined forces of the Sixth

U.S. Army under General Walter Krueger until Christmas 1944 to capture the entire island of Leyte in a slow and painful advance, it did not take the military commanders that long to realize that the airfields on Leyte were not sufficient enough to cover their forces in their push into Luzon—the northernmost and largest island of the Philippines where the capital Manila lies. The Allies needed better air cover, and Leyte's airfields were not up to the task. Three fast carrier groups had already begun closing in on Luzon on Bloody Sunday, and in the days that followed, air strikes began pounding the Japanese shipping in Manila Harbor.

Throughout the month of November, the kamikaze attacks continued with ferocity, as the Japanese perfected and improved their new type of aerial warfare with each passing day. The deadly effective tactic was proving to be extremely difficult to defend against. According to Morison's history, MacArthur himself had become so concerned over the threat that he ordered additional Hellcat fighters from Peleliu to Tacloban to protect them from air attacks.

The Allied command chose Lingayen Gulf as their landing target on Luzon, which would be a joint operation under MacArthur's command. On Leyte, the Sixth Army had encountered much stiffer resistance than they had anticipated, and compared to Leyte the Japanese had even larger numbers on Luzon. Therefore, the Lingayen Operation would have to be even more massive than the one at Leyte, which had set a record for the largest amphibious landing in the Pacific to date. Extra ships from Normandy would be needed.

However, things were not going as expected. The overly optimistic invasion date of December 20 for Luzon had to be scrapped, due to the fact that the Sixth Army had not completely secured Leyte Island whose airfields were also still not operating at full strength. MacArthur decided that the airfields on the island of Mindoro in the western Philippines were more suitable to their needs and must be seized before the final push into Luzon could begin. Its airfields also had the added benefit of being able to protect Allied shipping along the sea route from Leyte to Luzon, which made Mindoro an attractive and ideal target. Therefore, orders were issued for the seizure of

Mindoro Island on December 5, and the Lingayen Gulf operation was postponed until January 1945.

———

On November 15, while in San Pedro Bay, Stephen and the LCI 65 were moored along the starboard side of USS *Midas* (ARB-5), another landing craft repair ship that was converted from the LST 514. At long last, the 65 was finally getting needed repairs, as workmen and welders from the *Midas* had been working on the damaged gunboat since the day before. That afternoon, four of the 65's crew, including J.R. Reid's gunner and good buddy Odis Johnson, were under orders to depart the ship and board the LCI 68 with their baggage, service records, and pay accounts. The four lucky men would finally be transferring back to the United States. Stephen, J.R. Reid, and the rest of the crew said their farewells to Coxswain Odis Johnson, Boatswain's Mate 2/c Edward Wojciechowski, Signalman 2/c Rex Russel, and Radioman 1/c William Brown.

The extensive repairs lasted several more days. On November 17, Seaman 1/c Vern Durbin reported back aboard the 65 with his baggage. Stephen and the crew had not seen Durbin since the Morotai campaign when he evacuated in the middle of the night exactly two months earlier to seek medical treatment. On November 18 and 19, Gunner's Mate 2/c James Vincent, Seaman 1/c Henry Reeder, and Gunner's Mate 3/c Emory Young reported aboard the LCI 70, LCI 69, and LCI 68, respectively, for temporary duty. While they were gone, workmen from the *Midas* conducted additional repairs on the 65.

On the night of November 20, orders from the commander of LCI Flotilla Fifteen came down for the 65 to get underway and anchor in San Juanico Strait as quickly as possible, because there were reports that another typhoon was on its way. By 5:36 p.m., the gunboat was anchored in ten fathoms of water wedged in the Strait. At some point during the night, however, Seaman 1/c Robert L. Holland and Seaman 1/c Albert W. Mahan disappeared. By 10:25 a.m. the next morning, the two men still had not turned up, and Macaluso had no choice but to declare them AWOL (absence without official leave) at 1 p.m. when the crew was mustered. The LCI 64 came alongside to provide badly needed fuel, of which there never seemed to be enough. The next morning, November 22, Henry Reeder and James Vincent both briefly reported back aboard the 65, but then departed again that evening.

Stephen and the crew spent Thanksgiving 1944 as they had for the past month, anchored in San Pedro Bay, Leyte Gulf. But that battle-weary, exhausted, and always hungry men would not be having a fancy feast that year. In fact, the deck log of the ship has them recorded as being called to their battle stations for "flash reds" six times that day, and for the duration of November, the LCI 65 remained anchored and moored next to the *Midas* undergoing repairs and responding to anti-aircraft "flash reds" at their battle stations in the unbearably hot and muggy humidity of the Central Philippines.

On the last day of November, Stephen got some welcome news: his old friend from the LCI 329, Lt. (jg) Frank Love would be coming aboard the 65 with his baggage for temporary duty for the next few weeks.

18

DECEMBER 1944: A NEW SKIPPER

"The Leyte Operation made inordinate demands upon the troops. It is impossible for me to depict the hardships they had to endure or the desperate resistance they had to overcome."

– General Walter Krueger, Commanding, Sixth U.S. Army[1]

Over in Battlefield Europe a half a world away, a blistering cold winter was sweeping through the Ardennes Forest in Luxembourg—in the depths of the Nazi Rhineland. General Patton's Third U.S. Army was closing in on the Rhine River in their push into the heart of Germany. Army Private First-Class Oliver "Jerry" Budd, a butcher in civilian life and a Rifleman in the 5th Infantry Division, found himself trekking up a mountain in the frozen Ardennes with his regiment on a bitterly cold December day. A qualified marksman from Liberal, Kansas, Budd had always liked the Army.

Almost seventy years later, Budd described what happened that day. It was the moment the Battle of the Bulge—Hitler's massive last offensive in World War II—reached Budd and his unit. Once his 5th

Division had made their way to the top of the hilly terrain, they were ordered by their commanding officer to dig in.

"He told us to dig in, and that's what we did. When I started to dig in that's when I got hit." Budd remembered.

Little did the 5th Infantry Division know, they were being closely watched from across a nearby river. The Nazis were stalking the Americans from afar. They were watching their every move.

"They watched us closely." Budd soon found out. "They knew we were going up there. So they sent up bazookas, bombs, whatever."

A Nazi tank was positioned at the bottom of the hill unseen by the Americans.

"A tank was down below, shooting up." Budd recalled, although he didn't see it himself.

Right then an officer—positioned behind Budd when the Germans started firing—dropped. Budd saw the officer drop, but he did not drop with him. Budd later expressed his regret for not doing so. The tank fired up at Budd's position and exploded nearby his newly dug foxhole. Shrapnel from the German tank tore the left half of Budd's face off.

Suddenly, the officer began shouting to Budd from behind him when he noticed Budd was still alive on the ground, although missing half his face.

"[The medics] came up, put me on a stretcher, and took me to a base where I could be treated." Budd remembered. "[The officer] told me, after I was hit, 'Come back here with me!' So I did. I crawled back where he was. And that gave him time to call the medics. The medics got up there and got me, and brought me back," Budd recalled.

As suddenly as it had started, the war was over for Pfc. Oliver J. Budd. The officer's presence, as well as the medics, had saved Budd's life on that cold, snowy December day. Misinformation spread, however, back to his buddies on the line that Budd had died in the attack. Years later, when Budd and his wife Mildred reached out to an old friend, he was awestruck at discovering Budd was still alive after many years.

"It's all plastic surgery, a lot of plastic surgery. My jaws is locked, the roof of my mouth is gone. There's a seal up there (roof of mouth) so I can talk to you." Budd explained. "They sent me back to England . . . to be treated there."

Budd would require serious, long-term hospitalization that was unavailable so close to the front lines. He, like many who were severely wounded during the war in Europe, needed to get transported from the mainland to hospital ships farther out at sea, ultimately headed for England. One of the only ways to achieve this was by utilizing the different types of landing ships of the Amphibious Forces. LSTs and LCIs were among those amphibious vessels that played a crucial role in collecting and evacuating casualties from combat zones in both Europe and the Pacific.

When reflecting on the number of lives that were saved because of crews like his, Lieutenant R. Y. Martin, commanding officer of the medical LST 464, once wrote of his hospital ship:

> All hands worked tirelessly . . . Every man and officer did a remarkably fine job in defeating the enemy and reclaiming from death hundreds of American lives who would have perished except that these men arose from the dullness and monotony of war, and with inspired determination met the crisis![2]

Some men were lucky. Budd was one of those fortunate few who would make it home alive to tell his story. And coming home with him around this time were scores of excess LCIs, LSTs, and other ships from the Normandy landing that would eventually be bound for the Pacific Theater. Budd tenderly remembered that on his first liberty after arriving back in America, "My wife threw me a party when I came home."[3]

————

Back in San Pedro Bay, Leyte Gulf, Philippines, Stephen and the LCI 65 were all done with repairs from the *Midas* and were patrolling off Samar Island on December 1. LCI 69 pulled alongside the 65 that afternoon and Lieutenant Commander Archie Holmes came aboard for a ten-minute inspection. Shortly after, the LCI 70 came alongside the port side of the 65 to deliver some mail.

On December 2, the 65 entered drydock in the USS *Carter Hall* to undergo replacing their starboard screw. The crew spent the next two days painting the sides of their gunboat while the repair crews went about their work adjusting and repairing the damage to the 65's propeller.

Stephen would no longer have to worry about any issues with his skipper from Cleveland, Ohio, because on the afternoon of December 4, Charles J. Macaluso—who had played right tackle for the 1936 Notre Dame football team—mustered all hands at quarters and passed official command of the LCI 65 to the gunboat's new captain, Ensign William J. McKeon from Lansing, Michigan.[4] At 3:29 p.m., a small amphibious transport (DUKW) departed with Macaluso and the motor mac Edwards. Stephen never saw Macaluso again.

The 65 spent the next week continuing to receive repairs in Leyte Gulf, which included their radar equipment. On December 8, a small boat came alongside the 65 and finally picked up Walter Henry's gear, the young crewman who had been wounded and taken out of action at the invasion of Morotai in September. That afternoon, the USS *Rutilicus* (AK-113), a cargo ship came alongside to provide badly needed supplies and dry provisions. The next morning, the 65 moored alongside the LCIs 64 and 68 to transfer some of those supplies and provisions over to them.

On December 10, the 65 picked up two men from the LCI 69 and six men from the LCI 984 for temporary duty. Later that evening, the 65 took on over thirty-six hundred gallons of fresh water from the USS *Stag*, one of only four specialized water distilling ships built by the U.S. Navy during World War II. Ensign McCall, the 65's communications officer, took a working party aboard the *Stag* to gather more needed items. A few hours later, the men hurried to their battle stations due to a "flash red" but were given the "all clear" a half hour later. McCall and his men returned aboard with small stores shortly before 4:30 p.m. and the 65 anchored for the remainder of the night in San Pedro Bay.

The following week was spent doing more of the same in Leyte Gulf. The endless routine of pulling alongside various types of ships to take on water and supplies had become almost monotonous and mind-numbing to the exhausted, grimy crew of the 65. Stephen and his shipmates spent Christmas 1944 getting welding repairs for their gunboat alongside an auxiliary dry dock (USS ARL-8) in San Pedro Bay—the same place they'd been for over two months. They had spent the month of December gearing up for their next major invasion in the Philippines which was only days away: Lingayen Gulf. The island of Luzon would be Stephen's last invasion.

19

JANUARY 1945: LINGAYEN GULF

In serving on the seas, be a corpse saturated with water.
In serving on land, be a corpse covered with weeds.
In serving the sky, be a corpse that challenges the clouds.
Let us all die close by the side of our sovereign.

– Japanese kamikaze *"Song of the Warrior"*[1]

The location of Lingayen Gulf itself was a strategic advantage, so much so that the Japanese had used it when they invaded Luzon in 1941. Located in northwestern Luzon, Lingayen Gulf is a large, protected body of water that could fit nearly a thousand ships in it at once. It lies more than one hundred miles north of the Philippine capital of Manila and its low, level ground made for perfect beaching. Intelligence reports out of Luzon from the Philippine guerrillas at that time were claiming that the gulf was filled with enemy mines. Due to this, the Allies were to conduct mine sweeping operations three days prior to invasion day (dubbed S-Day) scheduled for January 9. Walter Krueger's Sixth Army would be conducting the ground invasion, and like the Leyte invasion, Admiral Oldendorf's six battleships would be leading the massive naval armada. Sixth Army Intelligence had estimated that over two hundred thousand Japanese troops awaited the Americans.[2] The Allies knew they were in for a bloody campaign.

———

"Luzon. That was the last push," an eighty-six-year-old Stephen remembered. He pointed to Lingayen Gulf on the map in front of him. "That's where Luzon Harbor is and that's where we went," he said.

On the morning of January 2, 1945, the LCI 65 formed up as the rear-most LCI in one column with the minesweeping convoy consisting of LCI (G) 69 as guide, followed by LCI (G)s 64, 68, and 70. As they departed San Pedro Bay, Leyte, each LCI kept a distance of about five hundred yards from each other in the column. In execution of Operational Plan 17-44, Task Force 77.2 was underway for Luzon. The convoy would be cutting through the Philippine archipelago as a shortcut instead of sailing all the way around the islands. LCI 70 veteran Gilbert Ortiz remembered, "We were in a fairly large convoy heading for the invasion of northern Luzon. We were in formation in columns of four."[3]

The convoy sailed undisturbed until around 7:00 p.m. that evening when the ships opened fire on two enemy bombers that began to attack the large ships in the center of the formation. The convoy managed to shoot one of the bombers down. At 4:45 a.m. on the morning of January 3, another lone enemy plane approached and dropped three bombs in the middle of the Allied convoy.[4] Then at 7:28 a.m., enemy planes began attacking the convoy again so the LCI 65 opened up with thousands of rounds from all of their combined guns. The 65 witnessed three of those Japanese planes shot down and destroyed. After that morning attack, the convoy continued sailing to their destination undisturbed for the next two days.

Then on January 5 (S-Day minus 4), all of that began to change. That morning, as the LCI 65 passed Manila Harbor following behind the small auxiliary minesweeper YMS-368, her crew went to battle stations three different times before noon. Several hours later, shortly after 3 p.m. the 65 received word from the commander of Task Group 77.6 that two Japanese destroyers had left Manila Harbor and been sighted only twenty-eight miles away from the convoy. As Stephen remembered it, "We passed by Manila Harbor about maybe fifty miles away from it, and pretty soon I looked up through the binoculars, and I told the old man, I said, 'There's two Jap destroyers coming on our ass.' So, he hollers out to the naval air force, you know, the carriers. And man, within a minute's time, they went right after them two destroyers and knocked them out." Reflecting on the potential danger of what could have happened, Stephen laughed as he added, "If not, they would've knocked us out!" But as Stephen put it, luckily "they knocked them out in a hell of a hurry."

Stephen moved his finger across the map along the path his convoy took as it sailed to Lingayen Gulf. He pointed to Surigao Strait at the south end of Leyte Gulf where that massive battle had taken place over two months earlier, then he began running his finger westward into the Sulu Sea, then he gestured north through the Mindanao Strait and up into the South China Sea.

As the fleet passed Manila Harbor at around 5 p.m., seven Japanese planes suddenly appeared in the sky approaching the convoy. The ships began to open fire on the enemy planes. That's when four of the planes broke off and headed straight for the LCIs. According to the LCI 64's records, the planes came in from their port side, which attacked the convoy individually, in order to take advantage of the sun low on the horizon.[5] This tactic of flying from out of the setting sun, as well as encountering heavy swells, greatly reduced the accuracy of the anti-aircraft guns.

The first plane, an A6M Zero fighter initiated a kamikaze attack on the convoy. The Japanese pilot descended into a dive at a forty-five-degree angle, crashing into the water about fifty to seventy-five yards off the LCI 64's starboard bow in a near hit. According to the LCI 64's Radioman 2/c Robert H. Barkan, "I was on the bridge, and I saw a plane start to dive on us, we gave him everything we had, he missed by 50 yards."[6]

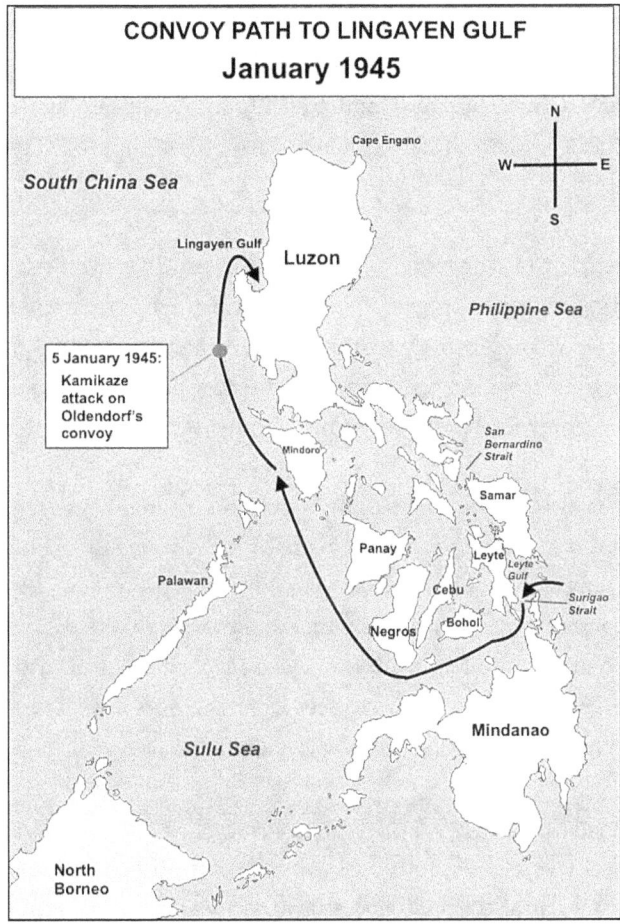

The Japanese attack on the Allied convoy on January 5, 1945. Modified version of original map: https://d-maps.com/carte.php?num_car=590&lang=en

The second kamikaze hit the ocean tug USS *Apache*. Atop the bridge of the LCI 64, Signalman 2/c James D. Robertson, remembered seeing this third kamikaze hit the *Apache*. In his book, he described the scene:

No sooner had we resumed battle readiness then we became involved in driving off another suicide attack. Our after guns offered up so much firepower the aircraft veered toward a seagoing tug maneuvering astern of the center column, USS Apache (AT-67), resulting in the loss of her radar mast and the wounding of three of the crew before exploding and crashing into the sea. On securing from battle stations note was taken: Apache took out four Japanese planes.[7]

According to Robertson, a third kamikaze, "like the first plane, . . . came in on our port quarter." This A6M Zero fighter made another dive for the LCI 64, but this time the kamikaze only missed the 64's forecastle by a few feet, crashing off the 64's port bow. However, this kamikaze fighter had been carrying a bomb. According to Robertson:

At the critical time, [the] skipper threw the rudder over "left full," reversed his port engines and called for "all ahead full" on his starboard engines causing the bow to turn sharply to the left. The plane, moving in excess of 350 miles an hour, unable to compensate for the ship's sudden maneuver, passed over the forecastle and crashed in the water about 20 feet from the port bow. The explosion caused bits and pieces of the plane's wreckage to rain down on us.[8]

When the bomb exploded as the kamikaze hit the water, it was so close to the 64 that the LCI was engulfed by water and debris. Robert Barkan described the feeling of relief after barely surviving such a close call after this second near hit saying, "Then another one came in, I . . . prayed and prayed. He missed us by just a few feet. Water came over the whole ship from the explosion. I thanked God for saving us."[9] It was after this second near hit that the crew of the LCI 64 officially coined her nickname—the "Lucky 64." This was now at least the fourth near hit of the LCI 64 by a kamikaze since the Philippines campaign began in late October.

The fourth kamikaze now approaching from the rear of the

convoy was the one that caused the most damage. Within a minute of the hit on *Apache*, the fourth Japanese plane dove for the LCI (G) 70.

———

Over on the LCI (G) 70, nineteen-year-old cook, Seaman 1/c Royal Wetzel, remembered at that very moment, "There was so much shooting going on." He recalled that once they got to Luzon on January 5, "That is where we got hit." At the time, Wetzel was at his battle station as the loader on the 20mm machine gun on the starboard side aft (back right of the ship) for his gunner, Frank L. Smith. According to Wetzel, "I was on the fantail of the ship and a kamikaze plane came in. I laid down flat on my face and said, 'Oh man, this is it.' I was lying on the deck and I saw him coming in. The kamikaze came screaming down at a steep angle from the rear of the gunboat and several sailors jumped off the ship before it crashed into them." Wetzel recalled, "I got up and some of the guys weren't on the fantail of the ship. They had jumped overboard . . . and one guy somehow or other threw his ring off before he went over. We found that on the deck. The plane came down and took the mast off and went right over the three-inch [50 caliber gun] and down it came onto the bow." He added, "When the plane hit, there was black smoke that came back over us."[10]

Gunner's Mate 3/c John A. Reulet had just celebrated his twenty-second birthday the day before. Now, Wetzel's shipmate and good buddy from Louisiana, was also at his battle station at another 20mm gun, amidship, right underneath the conning tower, on the starboard side. Reulet's gun was located in front of Wetzel's gun on the same side of the ship. Reulet remembered that the kamikaze "hit the conning tower and he cut the mast off . . . he must have been four or five feet above the conning tower. [He] [k]nocked the radar and everything down and he hit the bow . . . three boys we never found them." He said, "I don't know why [the kamikaze] picked on a little ship like that," but Reulet added, "We got hit there and we lost a few boys." Reulet would be awarded the Purple Heart from shrapnel wounds sustained in his leg from that attack. Despite one of his legs being seriously wounded, he insisted, "I was still on my gun standing on one leg..."[11]

A painting titled "The Unsinkable 70" depicts the kamikaze attack on January 5, 1945. Artist Joseph Ortiz dedicated this painting to his uncle, Gilbert Ortiz, of the LCI 70. (Reprinted with permission by Joseph Ortiz; also in the collection of the National Museum of the Pacific War in Fredericksburg, Texas)

Another crewman, Seaman 1/c Leo Wilcox from Los Angeles remembered the kamikaze attack and agreed with Wetzel's recollection saying, "Every gun that could bear on that plane was firing. I was firing at the plane with my twin 50 but one of the guns wasn't firing. I had a lot of trouble." Nineteen-year-old Wilcox kept cocking the gun, but it misfired instead. At that instant, Wilcox remembered the kamikaze "hit the mast about ten feet over my head and cut the mast off. Then he hit the bow in the middle of the ship." Wilcox recalled that the plane hit the gun tub where his buddy had been. "The plane hit directly on that gun tub and blew that bow completely off. We never found any part of the three gunners that were on that gun. One of the loaders was my best friend." Wilcox continued saying, "The gunner was Charles A. Poole. One of the loaders was Denzil Phillips and the other loader was a cook, George Pressley." When the plane hit, Wilcox said a bunch of guys abandoned ship. "They probably thought the ship was going to explode so they dove over the side."[12]

LCI 70 veteran Gilbert Ortiz in World War II. (Courtesy Joseph Ortiz)

It all happened very quickly according to Gilbert Ortiz, then a Seaman 1/c. "This is what I can remember about the kamikaze attack," he wrote in an email from 2008. "There were four zeros, each picked the last LCI on each column," he continued, "the [first] zero missed the LCI to our left," then, "a few seconds later we were hit. The plane came in from astern (rear) it clipped our mast before plunging into our bow (front) killing the [40mm] gun crew." Ortiz recalled that "the explosion also killed and wounded several of the crew in the 3inch 50 gun tub."[13]

The kamikaze attack on the LCI 70 left four men dead, two missing (later declared dead), and nine wounded. It caused eight men to jump overboard into the water to avoid the crash. The destroyer USS *Dorsey* eventually picked up five surviving crewmembers that jumped in the water. Reulet sadly remembered the *Dorsey* "came along before dark and picked up some of the boys that were wounded pretty bad, and three of them they didn't pick up because they were dying anyway."[14] According to Ortiz, "the men that were [killed in action] we buried them at sea" the next day.[15]

Crewmembers of the LCI 70 pose for a picture in Tijuana, Mexico after the war. (Left to right) Maurice "Paddy" Ryan (Cox.); Quentin C. Pearce (S1/c); Gordan W. Randall (S1/c); John A. Reulet (GM3/c); Royal A. Wetzel (S1/c). (Courtesy Royal Wetzel)

"Man, I felt so sorry for those guys," Stephen said of the sailors he witnessed jump overboard into the water from the LCI 70.

Commander Loud who was leading the minesweeping group—which included the *Apache*—came under attack by the remaining three kamikazes of the seven. All three crashed into the water, splashing close aboard three of the sweepers, but not damaging any of them. According to Morison's history, the radar screens did not justify an "all clear" until 6:40 p.m.[16]

Stephen's LCI 65 had expended over twenty-eight hundred rounds of ammunition by the end of that horrific kamikaze attack on January 5 that consisted of more than fifty suicide planes and had lasted for two and a half hours. The enemy damaged seven heavy Allied ships and the little gunboat LCI 70.[17]

This attack, however, was only just a prelude for the kamikaze attack that would be unleashed the next day.

On January 6, 1945, (S-Day minus 3) Admiral Oldendorf's battleships, cruisers, and destroyers, as well as the escort carriers, minesweepers, and demolition teams entered Lingayen Gulf at dawn to begin their three-day-long naval and air bombardment. The LCI 65, still just outside Lingayen Gulf, sailed in the convoy astern the LCI 69 at 7:45 a.m. The LCI gunboats of Task Unit 77.2.8 began to report-in to the commander of Task Group 77.3. As the heavy ships battered the beaches, the first Japanese kamikazes appeared in the sky shortly after 9 a.m. As Stephen stood at his battle station atop the conning tower, the 65 began firing all her guns. Once again, Admiral Barbey would be leading the Seventh 'Phibs aboard the flagship *Blue Ridge*. According to the officers' accounts of that morning who were present, the ships had hardly started their assigned duties of knocking out the coastal batteries, clearing out mines, and removing underwater obstacles when the kamikazes attacked their ships in a series of raids that lasted until sundown. Sixteen ships were hit and casualties were heavy.

By the end of the day, according to Morison's history, the Japanese kamikazes had unleashed their "most effective [attack] of the war in relation to the number of planes involved: 28 kamikazes and 15 fighter escorts."[18] But for Japan, these kamikaze victories were bittersweet. Each attack came at a great cost, as they were losing scores of their remaining pilots with each passing day. Dozens of the best trained Japanese pilots would eat breakfast each morning only to take off from airfields, never to return.

Later that brightly moonlit night, as Stephen and the crew of the 65 stood outside Lingayen Gulf astern the LCI 69, drastic measures were taken by Halsey's Third Fleet further north to intensify attacks on Japanese airfields on Luzon as a response to the kamikaze threat.

Then, on January 7, the Navy frogmen arrived. As the morning bombardment began at 10:30 a.m., and continued well into the afternoon, six underwater demolition teams (UDTs) began their work clearing mines and gathering intelligence ashore for the coming landings. Frogmen, as they were originally nicknamed by a war correspondent—equipped with masks and fins resembling frogs—were actually considered the first Navy SEALs. According to author Dick Camp, several days before the main invasions were scheduled to begin, these fearless divers would dive into the enemy waters "to gather intelligence—beach and surf conditions (obstacles, trafficability), Japanese defenses—and relay the information directly to the assault units." Camp added, "After completing the mission, they were to be taken by fast transports directly to the ships carrying the assault force to brief it on what they had seen and learned."[19] Ernie Pyle once remarked that these highly trained and courageous volunteers were "half fish and half nuts."

It would be Stephen's and LCI 65's job to protect these frogmen that day. At 7:00 a.m. the 65 entered Lingayen Gulf with the rest of the ships of Task Unit 77.2.8 and made their way toward the San Fabian landing area, which consisted of White Beaches 1, 2, and 3. By 2:00 p.m., the 65 was positioned off White Beach 1 to cover Underwater

Demolition Team Eight, that had made their way over from the USS *George E. Badger* (APD-33) in small boats. At exactly 3:15 p.m., with all ships in their proper pre-assigned positions, the 65 opened up fire on the enemy beach, bombarding it with all her batteries from five hundred yards offshore. Ten minutes later, the frogmen were placed in the water. Five minutes after that, the UDT officers requested the 65 cease firing her 20mm and .50 caliber guns. However, thirty minutes later, the 65 began bombarding the beach again. By 4:45 p.m., the 65 had completed her mission and ceased firing for good. She departed the bombardment zone and got underway, headed north in company with the LCI 69.

Without any rest, the men sailed north, where they stood and maintained their normal condition III watch while underway all through the night and into the next morning.[20] Around 9:30 a.m. on January 8, the 65 rendezvoused with LCI (G) 751 before joining the rest of Task Unit 77.2.8 in route to their next target, Poro Point— located off the city of San Fernando on the northeast coast of Lingayen Gulf. By 1 p.m., the 65 was in position off Poro Point astern the LCI 69 in the column. Shortly after, the 65 opened fire two thousand yards off the beach with their three-inch 50 caliber and 40mm guns on Japanese barges that sat beached on the shoreline. It only took ten minutes for the convoy to completely destroy the enemy barges.

———

"So anyway, we got up right here," eighty-six-year-old Stephen said as he motioned on the map to Lingayen Gulf. "Right in San Fernando, that town, we pulled in like this, and came right in this pocket." His finger moved south to San Fabian on the map. "And this is where our landing was," he said as he pointed to White Beaches 2 and 3.

On the morning of invasion, January 9, 1945 (S-Day), sixty-eight thousand troops of the Sixth U.S. Army stood off the landing beaches. There, they awaited H-Hour, where they would invade the largest and most heavily defended island of the Philippines. Ironically, for his last invasion, Stephen and the LCI 65 would be covering the landings of the familiar men from the 43rd Infantry Division— the same men Stephen had delivered to the beach in his very first invasion while serving aboard the LCI (L) 329 on Rendova Island back in July 1943. The LCI 65 would be tasked with covering White Beaches, specifically the right flank of White Beach 2, which was also the left flank of White Beach 3. The 172nd Infantry, 169th Infantry, and the 103rd Infantry Regiments were assigned to land on White Beaches 1, 2, and 3, respectively. The 6th Infantry Division, consisting of the 1st Infantry and 20th Infantry Regiments, would be landing at Blue Beaches 1 and 2, respectively.

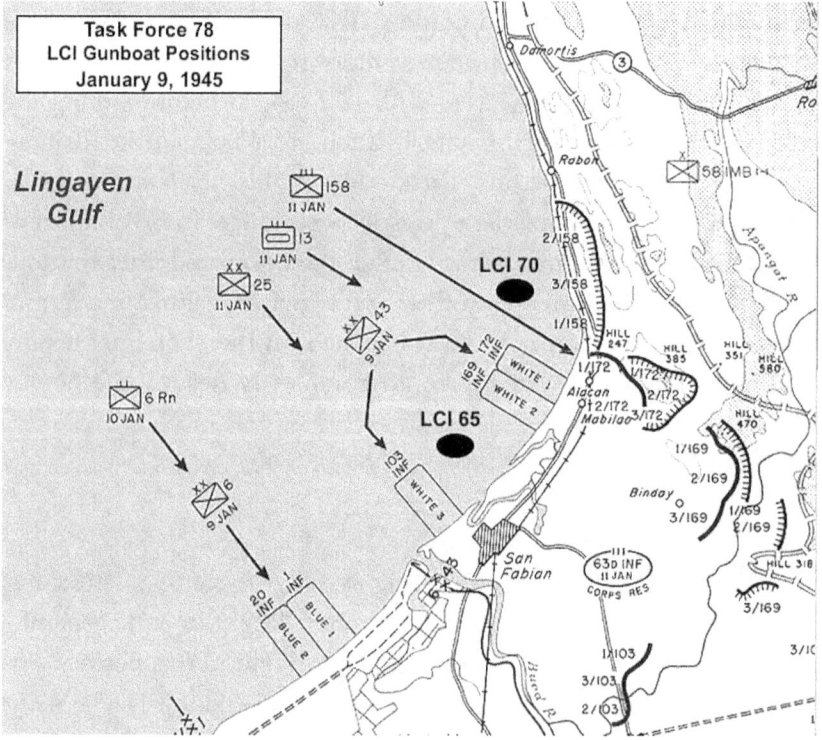

The assault at Lingayen Gulf by some of Task Force 78 and the positions of LCI 65 and LCI 70. (U.S. Army photo with labels added)

By 8:30 a.m., the LCI 65 approached White Beach 2 with the ships of Task Group 77.2 as part of Operational Plan 17-44. By some miracle, the LCI 70 was one of them, despite being severely wounded in the kamikaze attack four days earlier. It was determined that despite the damage inflicted, she could still participate in the landings on S-Day. The LCI 70 was assigned to cover the left flank of White Beach 1 and the left flank of the LCI rocket ships bombarding the beach. The LCI 65 arrived in her assigned position covering the right flank of White Beach 2 as well as the right flank of the column of rocket armed LCIs.

Stephen remembered the pre-invasion bombardment from the massive battleships as they were going toward the beach saying, "We had to pass the USS *California*. So, we were like, say maybe roughly five hundred yards—maybe eight hundred yards away from her. And she had her sixteen-inch rifles like that," he said as he extended both arms straight outwards, "and all of a sudden, she goes '*wah-boom!*' and you could see [water] like around the [*California*], like say for about three blocks, the water would just ripple, and that big wagon moves it three feet every time that big gun fired off. And you ought to have been underneath them. Boy, my bones hurt from the concussion!"

H-Hour began at 9:00 a.m. as the assault boats sailed past the line of departure and began approaching their designated beaches. When the assault craft were about twenty-five hundred yards off the beach, the heavy bombardment ceased. At 9:35 a.m., nine hundred yards off the beach, Stephen swung the ship right and—with her port side facing the beach—commenced firing all her guns on White Beach 2 as she moved westward toward White Beach 3.

By 9:46 a.m. the crew of the 65 spotted men ashore from the 103rd Infantry Regiment on White Beach 3 link up with remnants of the 169th Infantry Regiment that had landed on White Beach 2. As soon as this gap was closed, the 65 ceased firing. All day they remained at General Quarters off White Beach 2 until shortly after 5:30 p.m. They changed positions and anchored off White Beach 1.

An hour later, an unidentified plane came in low from behind the hills on White Beaches. It was identified on the bridge as a friendly P-61 Black Widow fighter, however, before the word could be passed to all hands, most of the ships in the bay opened up on it, as did four of the 65's gunners. Once they realized their mistake, the ships ceased firing immediately. The 65 remained anchored in the waters of Lingayen Gulf for the rest of the night and for the six days that followed.

———

Much has been written about LCI history and the sea stories from their sailors, but rarely has a personal account captured the essence of the life of an LCI sailor as eloquently as James Douglass Robertson did in his book, *Robby*, published in 2012. On Sunday, January 14, 1945 while Stephen and the LCI 65 were anchored in Lingayen Gulf, the gunboat LCI 64—now the Lucky 64—was on her way back to Leyte Gulf. The little gunboat and her battle-weary veteran crew would be among the first of Stephen's LCI Group Forty-Five of Flotilla Fifteen to get recalled back from the front lines in the coming days. On their way back to Leyte, the Lucky 64 spotted the injured baby flattop carrier USS *Salamaua*. The Lucky 64 had received medical supplies from the LST 68 earlier in the day, in order to come alongside and deliver blood plasma to the *Salamaua*. In poetic fashion, Robertson described their encounter in his book:

> None of us had ever seen a baby flattop. . . . Salamaua was a Casablanca Class. . . . She had taken a kamikaze, carrying two 550 pound bombs. Over 80 men were injured. Fifteen were killed. A witness to the attack observed: "The plane hit our ship in an almost vertical dive leaving the flight deck with a hole that looked almost like a silhouette of itself." . . . When she first hove in sight we all stopped in our tracks; this was a "baby flattop"? It was massive – a floating island; it boggled [t]he mind. We saw she was listing severely to the starboard and moving just fast enough to maintain steerage. Far, far up in the carrier's superstructure the signal bridge came to life. Their flashing signal light beamed down at us shooting a crisp series of dots and dashes which, when translated into words directed us to stand off and wait further orders.

Meanwhile, we approached from the stern being careful to maintain a parallel course of several hundred yards ultimately affording a full view of the carrier's port side. A burial ceremony honoring lost shipmates was in midservice on the center sponson. Strains of the Navy's Hymn – *O hear us when we cry to thee, for those in peril on the sea* –drifted across the clear blue water. As I carried out the order to "dip colors," I caught a glimpse of our 48 star, red, white and blue ensign streaming in a vivid blaze of color taking center stage for a brief moment catching everyone's attention. Time froze! A surreal panorama opened and we were part of it. A tiny ship going to the assistance of a crippled Leviathan. There was color, smoke, noise, motion, tragedy, and hope burnt into us in that brief moment never to be forgotten. We dipped our colors three times and left them at half-mast throughout the service.

The fallen, cloaked in traditional white canvas slipped silently from beneath the stars and stripes into the sea. . . .

Most of us, perhaps all of us, had never seen a carrier up close. The guys on the carrier similarly had never seen an amphibious vessel close up. Salamaua, even in this combat zone, was still cursed by a lot of spit and polish. Her hull was light gray and freshly painted. She had a place for everything and everything was in its place. There was no sign of extraneous gear piled up on her decks. It had all of the earmarks of a "ship of the line" although she was not. Those men not on watch or at alert stations were dressed in a modified uniform befitting the occasion. In this case: white hat, white T-shirt, white belt and uniform pants. From the looks of them they were fresh from a laundry other than a bucket, kiyi brush and a bar of yellow soap. Jealousy rose its ugly head thinking about their luxurious lifestyle. *How we envied them!*

In contrast, the LCI 64 was painted a peeling frog green. Our decks were strewn with all kinds of gear we had not yet found time or a place to stow. The crew was unshaven, dressed in our work "uniform" consisting of a pair of dungarees and work shoes and a dyed Navy blue hat. It was hot and everything, including us, stunk of diesel oil. The guys on the carrier lined up on the flight deck high above us and looked down on what they perceived to be a very casual lifestyle. *How they envied us!*[21]

It is worth noting that the burial ceremony held aboard the USS *Salamaua* was for Steward's Mate 1/c Dennison Vonner, who was killed in action by the kamikaze. Vonner was just one example of an African American man who, though relegated to the lowest job and facing rampant discrimination and hostility, gave his life for his country, and nonetheless found a way to become a shining example of courage for America.

———

That same evening, Stephen's LCI 65 departed Luzon and followed the Lucky 64's path as it limped back to Leyte Gulf under tow of the LST 636. Stephen's seventh and final campaign had been fought and he had managed to somehow survive yet again. He had seen the last of combat duty. Though he would remain in the Pacific Theater for another four months, his war was essentially over. The 65 would also need to undergo extensive repairs that would take months. It would take five days for the 65 to make it back to Leyte Gulf. The 65 would spend the remainder of January undergoing the first of their repairs in the familiar waters of San Pedro Bay.

20

FEBRUARY 1945: UNCOMMON VALOR

"The courage, effectiveness and fighting spirit of these small craft, lightly armed, vulnerable in construction, and manned by a mere handful of officers and men of brief naval experience, are not merely in keeping with our finest traditions and standards but add appreciably thereto."

– Fleet Admiral Chester W. Nimitz
Commander in Chief Pacific Fleet
(Action Report, August 13, 1945, detailing LCI action at Iwo Jima)[1]

On February 1, 1945, as the LCI 65 lay anchored in San Pedro Bay, Stephen was promoted to Quartermaster 2/c. It would be his final promotion in the U.S. Navy. The next day, his gunboat would begin its much-needed repairs from the USS *Midas*, and later that afternoon, after mooring alongside the LCI 69, the LCI Group commander Archie Holmes came aboard for transportation to Leyte Island. After dropping him off an hour later, the 65 spent the rest of the evening taking on fresh water from the USS *Stag*.

Finally, on February 6, Stephen and the crew of the 65 got some welcome news. After more than three months of non-stop action in the Philippines area of operations, the crew would finally be headed back to New Guinea to rest and receive other repairs. At 2:00 p.m. sharp, the LCI 65 cast off all lines and got underway with the LCI 69 on their journey back to New Guinea.

On February 11, the convoy made up of LCIs 65, 64, 68, 69, and 70 arrived off the southern entrance of Woendi Island, New Guinea. The crew got to work right away, and they spent the remainder of February alongside "Liberty Wharf" off Woendi Island undergoing repairs and resupplying what seemed like a completely empty stockpile. The last three months had put quite an amount of wear and tear on the little gunboat.

———

On February 17, 1945, an event happened that would never be forgotten by LCI veterans of the Amphibious Force, regardless of the theater in which they served. That day would go down in history as the single deadliest day for LCIs, which includes the D-Day invasion at Normandy. As Stephen was moored on the LCI 65 in the waters off Woendi Island, New Guinea, a legendary battle had occurred in the early morning hours a little over two thousand miles to their north. The Central Pacific Force of the U.S. Navy under the command of Admiral Chester Nimitz was scheduled to conduct their underwater demolition operations off a tiny island called Iwo Jima.

Many have heard the countless tales of "uncommon valor" displayed from the main invasion of Iwo Jima that began on D-Day (February 19) by the 3rd, 4th, and 5th Marine Divisions. Many more have seen the famous American flag-raising photograph taken atop Mount Suribachi by photographer Joe Rosenthal on February 23. But few have heard the incredible story that occurred two days before the invasion (D-Day minus 2) off the coast of Iwo Jima. What happened there to those little LCI gunboats of Flotilla Three as they assisted the Navy frogmen was unthinkably tragic. That day, ten LCI (G)s supporting the UDTs suffered a total of 201 casualties (30%). Of this number, 47 (7%) were fatalities.[2] One of the LCIs that fought against the Japanese that day was the LCI (G) 449.

The unforgettable story of the LCI 449, too long for inclusion in this book, was documented in meticulous and splendid detail in Mitch Weiss's 2016 book titled, *The Heart of Hell: The Untold Story of Courage and Sacrifice in the Shadow of Iwo Jima*. This book is highly recommended for any person interested in learning more about those valiant LCIs of Flotilla Three.

The famous flag raising photo taken atop Mount Suribachi, Iwo Jima, by Joe Rosenthal of the Associated Press on February 23, 1945.

Aboard his LCI mortar boat, radioman Garnett Ridenhour sat in the radio room of the LCI (M) 638. Four days after D-Day, Seaman 1/c Ridenhour was located several miles off the coast of Iwo Jima when he suddenly received a curious message. As Ridenhour remembered it, "Iwo Jima was about five miles long and a mile or so wide with

Mount Suribachi (dubbed 'Hot Rocks') an extinct volcano at the south end. . . . About the fifth day we were at the south end of the island in the vicinity of Hot Rocks. I took a message that said, 'Old Glory has just been raised on Hot Rocks.' When I had a chance to go topside I could see the flag flying at the top of the mountain. What a thrill!"[3]

21

MARCH–AUGUST 1945: END OF THE WAR

"We thought the Japanese would never surrender. Many refused to believe it. Sitting in stunned silence, we remembered our dead. So many dead. So many maimed. So many bright futures consigned to the ashes of the past. So many dreams lost in the madness that engulfed us. Except for a few widely scattered shouts of joy, the survivors of the abyss sat hollow-eyed and silent, trying to comprehend a world without war."

– Eugene B. Sledge, *With the Old Breed at Peleliu and Okinawa*[1]

Stephen Ganzberger poses for a photo on Woendi Island, New Guinea, March 1945.

At this point in the war Stephen was suffering from debilitating insomnia, as well as a host of other health issues including constipation and anxiety. He had been having horrible nightmares for months and hardly slept. More than two years aboard the tiny LCI gunboat barely eating, sleeping, and having regular bowel movements had taken its toll. The trauma of witnessing one of his best friends, Aiston, die in front of him was also contributing to his inability to sleep. It seemed to Stephen like he was always irritated and easily provoked to anger. He no longer had a desire to engage with his buddies. He had hardly been eating in the last several months. Most of those two years, Stephen was either engaged in combat, on watch, or was constantly on the move. As has been described throughout this book, life aboard an LCI was not easy. Stephen's lack of sleep was noticeably causing his health conditions to worsen. Others around him began to notice as well, including the officers.

After spending most of the month of March at Woendi Island undergoing basic repairs, the LCI 65 entered drydock for even more extensive repairs on March 20. Three days later, Ensign Cline, E. Harrison, R. Ellis, and Stephen left the ship to go to Biak on official business for two days. On either March 23, 24, or 25, Stephen was stopped on Woendi Island by a photographer who offered to take his picture. Stephen made his way over to a nearby tree and posed. After all he had been through, he found it in himself to crack a smile. The unknown photographer snapped Stephen's photo and after receiving a copy, Stephen grabbed a pen and wrote "Love Stevie" on the bottom right-hand corner of the photo. He put it in an envelope and sent it to his older sister Ida back home in Michigan.

———

"On April 1st, I was amazed at the huge, and I mean huge, assembly of all types of Navy ships," LCI (R) 706 veteran W. Donald Stewart wrote of the Okinawa invasion. More than two thousand miles north of Stephen's position at Woendi, Electrician's Mate 1/c Stewart was helping load rockets into the launchers that provided covering fire for the troops landing on L-Day. Okinawa—an island just over four hundred miles from Kyushu, Japan—was about to become the site of the last major battle of World War II and the largest sea-land-air battle in history. Stewart remembered, "On April 1, 1945, all fun and games were over, we ran within 100 yards of the beach and for several minutes, fired 36 5-inch rockets. . . . By midafternoon, all crew, who at different times participated in loading rockets into the launchers, became so tired they had difficulty getting 12 much less 36 rockets, into the launchers. These rockets weighed about 75 pounds a piece and had to be carried . . . up a couple of flights to the gun deck."[2]

Official records state eighty-one LCIs participated in the L-Day landings at Okinawa. But in the days ahead, even more LCIs participated in smokescreen and picket line duties as the kamikaze attacks began to rain down in a "typhoon of steel." The LCIs also simultaneously fought off suicide swimmers from nearby land.

"Japanese suicide boats, . . . loaded with explosives, would try to crash into one of our larger ships," Stewart recalled. The vitally important LCIs were in constant danger throughout the entire Battle of Okinawa—a battle that would claim over fifty thousand American casualties including over twelve thousand killed. Eugene B. Sledge, a Marine who fought on Okinawa in the 1st Marine Division wrote of the nightmarish battlefield years later, "We were in the depths of the abyss, the ultimate horror of war. . . . We were surrounded by maggots and decay. Men struggled and fought and bled in an environment so degrading I believed we had been flung into Hell's own cesspool."[3] Still more unsettling was that the Allies were noticing the enemy fought harder and harder the closer they pushed toward the Japanese mainland.

But perhaps the most heartbreaking news out of the campaign was that of the death of icon Ernie Pyle, who was shot and killed by a Japanese sniper on the nearby island of Ie Shima. News quickly spread to the front lines that the world had lost one of its most beloved, dedicated, and well-known American war correspondents. America was reeling after a one-two gut punch.

———

In John Dingell's words, "I had barely started Officer candidate school at Fort Benning, Georgia, on April 12, 1945, when President Franklin Delano Roosevelt died suddenly."

Roosevelt had been seated in a chair in front of a fireplace in the Little White House atop Pine Mountain in Warm Springs, Georgia. An artist had been laboring away making sketches of him. The president had been in excellent spirits since 9:30 that morning. That all changed when Roosevelt felt a sharp pain in the back of his head.

"I have a terrific headache," Roosevelt announced abruptly at around 1:15 p.m. Eastern Time. For the incredible man who ushered in the New Deal, helped pull America out of the Great Depression, and whose voice had brought comfort and security using his fireside chats—those five words would be his last. Roosevelt lost consciousness soon after and never regained it. At 5:58 p.m., the White House announced that the former Democratic governor of New York, and thirty-second president of the United States had died of a brain hemorrhage on the eighty-third day of his fourth term at age sixty-three. He was the only U.S. president to have been elected to serve more than two terms. According to Dingell, "America as we had known it for more than a dozen years changed completely in a single instant. For most people my age and younger, FDR was the only president we'd ever known."

Dingell had personally witnessed the president's famous speech in the Capitol the day after the Pearl Harbor attack. Heartbroken, he remembered that he walked around in a daze for hours following the news, but then found his bunk. Shortly after he began cleaning his rifle, he was approached by one of the men in his platoon. In his book, Dingell wrote:

> I looked up to see what seemed like the whole blasted platoon gathered at the foot of my bed! One of them asked me, "Dingell, your dad knew Roosevelt. Did you know him?"
>
> I didn't want any special treatment in the army, so I asked my

dad not to write me using his congressional letterhead. And I didn't want any special breaks, so I never told anyone that my father was a congressman who was good friends with our commander in chief. How in the bare-ass hell did they *all* know?"[4]

But that question no longer mattered. The question that did matter was the one on all the men's minds, including Dingell's. What kind of a fellow was this new guy Harry Truman? Dingell was curious about this new president as well, so he wrote his father to find out. About a week later a letter from his father came back with his official congressional letterhead. In his letter, John's father assured his son and the men of his platoon to have no doubts about their new president. According to John Dingell Sr., Harry Truman was a great man that had the "courage of a lion." After reading his father's letter aloud to his buddies, Dingell wrote, "A few of them patted me on the back. Most of them were grinning. They felt good again. Now they were sure we had a new president who was as strong as Roosevelt and who was going to win this effing war—even if they didn't know my dad." Dingell remembered that it didn't take long for news to spread, and that word all over the battalion was, *"It's gonna to be all right! Dingell says so."*[5]

———

News of Roosevelt's death reached Stephen and the crew of the LCI 65 while they were undergoing repairs at Woendi. The sailors who served aboard the LCIs were devastated. According to quartermaster Joe Harris of the LCI 600, "April 12, we heard the news while on the train, that President Roosevelt had died. An old lady sitting near me started crying and I told her that things would work out, that she shouldn't worry."[6] Royal Wetzel of the LCI 70 added, "Everybody was down in the dumps then. Everybody thought [Roosevelt] was a pretty good guy."[7] Harold Ronson of the LCI 1012 remembered the exact day FDR died, because it was his and his father's birthday. Ronson wrote that Roosevelt "was our commander-in-chief and a bit of us died with him that day."[8]

In early May, the crew of the 65 received news that the war in Europe was over. Days prior, from within his underground bunker in Berlin, Adolf Hitler committed suicide as the Soviet Red Army approached. Nazi Germany had finally been defeated by the Allies. The 65, however, was still at war and was undergoing repairs and resupplying for their next major invasion slated for June at Brunei Bay, Borneo. Stephen, however, would not participate in that campaign. On May 15, 1945, while moored at Morotai Island, Stephen was given orders by the skipper McKeon to transfer to a Naval Base Hospital for further medical treatment and diagnosis. At 1:00 p.m. a small boat came alongside the 65 to take Stephen away from the gunboat for good. Stephen said goodbye to all his buddies that still remained on the ship. He and a pharmacist's mate named Coleman departed in the small boat. But before sending Stephen off, McKeon included in his paperwork a recommendation for service school that he typed up that morning. In his recommendation, McKeon wrote the following:

1. As of this date subject man has served twenty-seven (27) months continuous duty beyond the continental limits of the United States. Most of this time has been in the most forward areas without adequate facilities for rest and recreation.

2. For the past several months GANZBERGER has been senior petty officer in the "C" division aboard this vessel and has executed his duties in the most satisfactory manner.

3. Therefore it is recommended by this command that upon his return and reassignment in the United States that GANZBERGER be given every consideration for Advanced Quartermasters School if there be a quota open. It is the feeling of this command that such an assignment would not only be advantageous to the Navy but would

also give the man a well-earned period of rest after an extended tour of duty in the Pacific as noted in paragraph one.

[Signed]
    W.J. McKeon Jr., Lt.(jg), USNR

Stephen had been aboard a Landing Craft Infantry for more than two years, but now his war was finally over. After saying his farewells and departing he took one last look at the 65. He would never see his ship, McKeon, or any of his buddies ever again.

On May 21 Stephen arrived at U.S. Naval Base Hospital No. 16 on Manus Island in the Admiralty Islands. He spent the next several weeks undergoing medical examinations and getting some much-needed sleep before he was ordered to be sent back to the United States on June 7, the same day the 65 arrived at Borneo for their invasion. Stephen was finally going home to the States.

Stephen arrived back at U.S. Naval Hospital in Great Lakes, Illinois, on July 3 after a brief stay at a naval hospital in San Francisco. Now a Quartermaster 2/c and Senior Petty Officer, he had returned to the place where he underwent boot camp as an Apprentice Seaman nearly three years earlier.

After four weeks of examinations, the doctors who conducted the medical surveys diagnosed Stephen with a case of "operational fatigue" due to twenty-eight straight months of duty in over six major engagements. Operational fatigue, which they also called psychoneurosis anxiety, used to be the medical term for what we refer to today as Post Traumatic Stress Disorder (PTSD). Among Stephen's most troubling symptoms were insomnia, catastrophic battle dreams, nervousness, and anxiety. Below is the full diagnosis by the Great Lakes medical staff:

This 20 year old QM2c with 2 years 10 months and 1 day of active duty prior to this admission to the sick list was readmitted to this hospital on 5 July 1945 under the diagnosis of Operational Fatigue complaining of nervousness and fatigue. This man has served 28 months overseas having participated in six major engagements. According to the patient's accepted statement and his health record entries he has been suffering from his present symptoms during the past year and particularly since his ship was hit by a crash dive in which his buddy was killed. He complains of nervousness, headaches, fatigue, profuse sweating, chronic constipation, some anorexia, irritability, restlessness and inability to tolerate routine and regimentation or get along with his shipmates. This condition has gradually increased in intensity since its onset and continues to date. It interferes with his performing his duty adequately and has resulted in his being evacuated from overseas.

Past history shows that this patient made a normal, adequate adjustment scholastically, industrially, and socially prior to enlistment.

Physical, neurological and routine laboratory studies are essentially negative. Psychiatric examination reveals a nervous apprehensive, restless but cooperative individual who is preoccupied with a fear of further sea duty and thoughts concerning his future security and health. His stream of speech is clear and relevant. The mood presented seems normal. His tension state is manifested by digital tremors, palmar sweating, and startle[d] reactions.

It is the opinion of the Board of Medical Survey that this man suffers from psychoneurosis Anxiety which renders him unfit for further duty. It is recommended that he be discharged from the U.S. Naval Service. At present he is in no further need of hospitalization, does not present a menace to himself nor to others, is fully competent to be placed in his own custody and is not likely to become a public charge.

[Signed]
    C.L. REYNOLDS, Lt.Comdr. (MC)
    L.E. MANDERMACK, Lieut. (MC)
    F.P. MILLEN, Lieut. (jg) (MC)[9]

Meanwhile, on the other side of the world, certain events were unfolding that would ensure the medical staff at Great Lakes Naval Training Station would never finish the process of discharging Stephen from the navy due to his medical issues. On July 26, 1945, with the war in Europe over, America's new president Harry Truman gathered with the Allied leaders in Potsdam, Germany, to make a renewed demand for Japan's unconditional surrender. The ultimatum was named the Potsdam Declaration. But this ultimatum was different than previous calls for Japanese surrender. This one contained dire warnings that should Japan fail to put down its weapons, their home islands would be attacked, and they would be completely destroyed. As was always the case, Japan ignored the ultimatum and the war in the Pacific raged on.

———

In the early morning hours of August 6, 1945, an American B-29 Superfortress took off from a runway on Tinian Island in the Northern Mariana Islands and headed north toward Japan on a secret mission. The B-29 known as *Enola Gay* was piloted by Colonel Paul Tibbets and was destined for a city on the Japanese home islands called Hiroshima. Just as the sun was rising, the *Enola Gay* rose to its bombing altitude of 31,060 feet. Upon reaching Hiroshima, the B-29 dropped an atomic bomb named "Little Boy" at 8:15 a.m., which vaporized nearly the entire city below, killing tens of thousands of Japanese, many of them civilians. The explosion from the blast looked like God Himself had ripped open the sky exposing everyone to a fireball so bright, it surpassed the intensity of the sun. Three days later, on August 9, another American bomber repeated the same action but over the Japanese city of Nagasaki. On August 15, 1945, Emperor Hirohito officially announced Japan's unconditional surrender to his people over the radio. It was the first time the Japanese people had ever heard Hirohito's voice. The most violent and devastating conflict in human history was finally over. V-J Day celebrations broke out all over the United States and the rest of the world. Later that same day back in Illinois, Stephen was honorably discharged from the Navy.

Five days earlier, on August 10, John Dingell Jr. had been commissioned as a second lieutenant in the Army. Upon graduating from Infantry OCS, he was informed that he would have been in the first wave of the invasion of mainland Japan planned for November 1945. Dingell reflected in his book that "President Truman's decision to use the atomic bomb to end the war almost certainly saved my life."[10]

Robert Rosenwald, the sailor who had watched the Southern Cross constellation appear in the sky aboard the LCI (L) 1008 on his way to Bora Bora at the beginning of the war, lived to celebrate V-J Day and never forgot the feeling of witnessing his favorite northern hemisphere constellation reappear on the journey home:

> I can still remember on that trip from the [Philippine Islands] to Pearl [Harbor] when I first spotted the Big Dipper. It was just barely above the horizon. I turned to the Bos'n and said, "Fleck, look at that Big Dipper. We are heading for home. Let's celebrate!" He went below and brought up hot coffee and a sandwich and we celebrated. Each night that Big Dipper got higher and higher in the sky. What a warm feeling."[11]

———

After World War II, Stephen's second ship, USS LCI (G) 65, was decommissioned, struck from the Naval Register, and sold on February 7, 1947—its fate unknown. Stephen's first ship, USS LCI (L)

329, was used as a target for the atomic bomb tests conducted at Bikini Atoll in the Marshall Islands, known as "Operation Crossroads," in July 1946. The LCI 329 was scuttled off Kwajalein Atoll on March 16, 1948—its exact fate is also unknown.[12]

According to LCI 74 veteran Robert Kirsch, "Only twenty-one LCIs were sunk by enemy action," in the Second World War, "and many more were shot up, bombed, torpedoed, struck underwater mines, were hit by kamikaze planes," but nevertheless survived. He remembered stories of "LCIs returning from an invasion with so many bullet holes in the hull that they looked like a sieve." Yet, even with all they were forced to overcome throughout the war, LCIs "proved to be quite seaworthy."[13]

Immediately following the war, some LCIs served as part of the occupation force of Japan. Both the U.S. and Royal Navies largely phased out LCIs after World War II, however, several LCIs were used during the Korean War. Many other LCIs were converted to specialized auxiliaries. Over the years, most LCIs were decommissioned and scrapped, including most of the ones turned over to other countries' navies. There are hardly any former LCIs still in commercial use, but the ones that are have been much altered. A number of other LCIs were privately purchased and used as lighters, inshore cargo vessels, vehicle and passenger ferries, sightseeing boats, and even a floating restaurant. Two LCIs from World War II remain on display today: the LCI 713 in Portland, Oregon, and the LCI 1091 in Humboldt Bay, California.[14]

For those interested in learning more about the history of LCIs and the veterans who served aboard them, their official veterans' association, the USS Landing Craft Infantry National Association, is a great resource. The National Museum of the Pacific War in Fredericksburg, Texas, is another great resource, as it contains the entire collection of LCI records and documents that was donated to the museum by the USS LCI National Association back in March 2013.

22

HOME

"As I look back I marvel at the initiative, willingness, cooperative good nature, courage, and common sense of the average American far from home. Thrust into a strange environment, he was to risk his life in a war of which he had little knowledge and less understanding."

– Admiral Daniel E. Barbey[1]

Stephen Ganzberger (right) with his wife, Patricia Ganzberger—who was
Michigan's 15th District GOP chairwoman at the time. They share a laugh with
Stephen Bonczek, city manager for East Detroit. Image used with permission of
MediaNews Group. All rights reserved. (*The Mellus Newspapers*, December 19,
1984)

Stephen and his three brothers survived the war. Upon returning to
civilian life in Southgate, Michigan, Stephen worked as a sheet metal,
heating, and air conditioning mechanic for more than fifty years.
After several years he married his wife Patricia Ganzberger on April
28, 1951, and they remained together for sixty years. They eventually
had six children: Stephen, Michael, Patrick, Heidi, Victoria, and
Barrie.

John Dingell Jr. was discharged from the Army on November 13,
1946. On September 19, 1955, his father, Congressman John D. Dingell
Sr., who had spent his life dedicated to public service, passed away.

The elder Dingell's son, "Artillery John," was elected to represent Michigan's 15th Congressional District to replace his father later that same year. From that moment on, John Dingell Jr. would begin his fifty-nine-year tenure as the longest-serving congressman in American history until his retirement from the U.S. House of Representatives in 2015.

Stephen was elected and served as a Southgate City Councilman for fourteen years from 1970 to 1984, in Southgate, Michigan. He was even elected as city council president in the late 1970s. His wife Patricia would go on to serve as Michigan's 15th District GOP chairwoman in the early 1980s. It is through these political connections that Stephen and Congressman Dingell met and became close friends. John's wife, Debbie, also formed a lifelong friendship with Patricia. Stephen and John always shared a special bond as veterans of World War II and as scrupulous men of principle. In 2009, John Dingell personally wrote the Department of the Navy to assist Stephen with obtaining replacement medals he never received. A Navy official rectified the issue by providing Stephen's additional awards and added in a response letter, "I have also enclosed the silver star and the two bronze stars appurtenances that were not previously issued for the Asiatic Pacific Campaign Medal and Philippine Liberation Medal."[2]

Stephen was also elected as a credit union president, and later as president of Sheet Metal Workers Local 80. He was instrumental in many community and fundraising projects, and also served as the employee representative to the Civil Service Commission. When he retired, he became a licensed builder, as well as a heating, ventilating, and air conditioning contractor. He also served as liaison to the Capital Improvements Commission, where he provided input into the construction of City Hall, the police and fire departments, and the courthouse. In his free time, his passion was playing golf—his favorite sport. For the rest of his days, he always kept devoutly close to his faith, regularly attending St. Pious Church in Southgate with Patricia on Sunday mornings. For as much as he loved playing golf, he loved and cherished his family most of all.

Stephen Ganzberger (left) and Congressman John D. Dingell Jr. This photo was taken in 1992 at the wedding of Stephen's daughter, Vicki. (Courtesy Vicki Scarcella)

Stephen and his daughter, Heidi, at the Heritage Days Parade in 1977. Heidi was
crowned homecoming queen in 1976. (Ganzberger Family Collection)

"The crusty, old Regular Naval Officers of World War II had a very
low opinion indeed of our Reserves, Draftees, and Volunteers,"
retired Admiral John H. Morrill wrote decades later. But Morrill knew

better; he had witnessed firsthand the capabilities of LCI sailors. When describing his men of the LCIs who helped win World War II, Morrill declared, "Our officers and men had become Cincinnati—citizen soldier, sailors. Like the ancient Cincinnatus, they were ready to go and fight for the defense of their native land."[3]

In his book, James Robertson of the Lucky 64 recalled the pure disgust he initially felt the first time he laid eyes on the LCI 64, writing, "My first thought was along the lines of, 'What a piece of trash. Where did they ever find enough scrap iron to put this sorry rust bucket together?'" Robertson wrote he even begged his skipper for a transfer "less than a half hour after my arrival aboard," but despite the unpleasant first impression, he slowly grew fond of the LCI. Sometime afterward, while eating with his shipmate Robert Barkan, Robertson remembered turning to him and remarking, "'Someday we'll look back on this as the best days of our lives,'" at which point he remembered they "laughed hysterically at what we considered to be the stupidest thing I ever said—completely oblivious to the truth of the matter."[4]

One of the things the LCI veterans interviewed for this book have in common is nearly all of them shared Robertson's sentiment about LCIs. In the words of Dr. R. William Clark, "We developed a sort of love for these hard-working little ships and used them to win WWII."[5] Many, like Robertson, will admit that life was indeed difficult aboard those small, lousy LCIs but they nonetheless were extremely proud to have served their nation and their buddies that fought alongside them and found moments to enjoy their wartime naval service. Others, like Stephen, mostly reflected on their memories of the war in serious tones remembering valuable lessons learned and close buddies lost. Of those, plenty struggled with the fact that they returned home when many others did not. Patricia Rone, wife of LCI 683 veteran Gerald Rone, described her husband's pain in 2007 writing, "Even as his memory fails him, still there are times of that period, triggered by what I do not know, which reduces him to tears for just a few moments and then he is alright again. I am sure you have heard such many times."[6]

But sometimes, every once in a while, when the LCI veteran "old salts"—these brave Cincinnati now in their nineties—gather together at reunions and swap sea stories, especially late at night, one can tell by the looks on their faces and the joy in their laughter when they're reunited with their shipmates, that despite the horrors of war that were all around, the time spent on those tiny ships with their buddies, for many of them, truly was the best days of their lives.

"In the end," Robertson's assignment to an LCI was in his words, "the very best decision I never got to make."

———

Stephen's friend, John Dingell Jr.—the longest serving congressman in United States history and Presidential Medal of Freedom winner—passed away on February 7, 2019. His wife, Debbie Dingell, still serves as the U.S. House Representative for Michigan's 12th District as this book is written.[7]

On a sunny afternoon in April 2016, loved ones and friends began receiving the news that LCI 502 veteran John Cummer had passed away peacefully in his home in South Carolina. He had just finished breakfast and decided to take an afternoon nap. But he never awoke. He was ninety-one years old, which is ironic and fitting in a way, because his favorite Bible verse that he lived by was *Psalm 91*. The same *Psalm 91* that he had read and reread to himself over and over from within the LCI 502's dark, hidden compartment, the night before the Normandy invasion. The former LCI National Association president was a gifted writer and will always be an inspiration to those who knew him.

All the LCI veterans from World War II who were interviewed for this book have since passed away, except for one: Desmond Johnson of the LCI 69, who currently lives in Naples, Florida. As this book was going to print, Royal Wetzel of the LCI 70 died on October 31, 2021. Wetzel—aka The King of Kazoos—had a heart of gold and a face that resembled Popeye. He was a regular attendee of the annual LCI National Association reunions and became a sort of mascot for the association due to the joy he could bring just by being around. The cook from the LCI 70 who had survived a kamikaze attack at Luzon knew exactly how to make his friends and loved ones smile. He was the first and only LCI veteran to read a manuscript copy of this book before it was published.

On the night of May 20, 2011, two days after he shared his extensive World War II story in a two-hour interview with his grandson, Stephen Ganzberger passed away at Henry Ford Wyandotte General Hospital while surrounded by his children and grandchildren. His wife Patricia later passed away on March 7, 2017. They are survived by six children, seven grandchildren, and five great-grandchildren.

Stephen Ganzberger is buried nearby his friend John Dingell in Arlington National Cemetery.

Stephen and Patricia Ganzberger in 2008. (Ganzberger Family Collection)

EPILOGUE

"No other Navy at any time has done so much. For your part in these achievements you deserve to be proud as long as you live. The Nation which you served at a time of crisis will remember you with gratitude."

– James Forrestal, Secretary of the Navy (1946)

"They took the cream of the crop and killed 'em," Stephen once said. "A lot of kids got killed there."

World War II was the largest, bloodiest, deadliest, and most chaotic conflict in human history. An entire generation of men from all over the United States came together to defeat two of the evilest forms of fascism that have ever existed. The extraordinary victory the Allies achieved simply could not have been achieved without the ships of the Amphibious Force and the young crews who manned them. They became the pride of America during the Second World War and what they accomplished will always be worth remembering.

To conclude this epic chronicle of such valor, bravery, and patriotic duty of the men of the Landing Craft Infantry, it seems most appropriate to end with my favorite LCI exchange from the war that took place between Admiral Harry W. Hill and Commander Theodore Blanchard. Blanchard oversaw the LCIs' smokescreen operations at Eniwetok Atoll as they covered a group of cruisers and battleships under Admiral Hill's command during the Marshall

Islands campaign in February 1944. The exchange was first published in an article by United Press staff correspondent Richard W. Johnston, and republished in the February 2013 *Elsie Item* newsletter. The conversation went as follows:

> **Admiral Hill:** (to Blanchard) *The gallantry and contemptuous disregard for danger displayed by all LCIs attached to this command has been an inspiration to us all.*

> **Commander Blanchard:** (reply) *We are deeply grateful for the unselfish and gallant assistance your officers rendered. The cooperation and courtesy tendered to all LCIs by the cruisers and battleships has made us all feel that we too, small as we are, are a part of a great fleet.*

Johnston concluded by writing, "Perhaps the heretofore anonymous heroes of the LCIs needed Admiral Hill's reassurance to feel 'part of a great fleet.' You may be sure the Japanese had no doubt of it."[1]

———

As my grandfather lay on his hospital deathbed on the night of May 20, 2011, I held his hand one last time, and promised him that I would never let people forget about the sacrifices he made and the things he once did for this country. This book's purpose is to keep that promise. My hope is that after this book is published it will be passed down to those in my family who come after us. A very special thank you to all who read this book. I am eternally grateful to all those who contributed to its creation. To those readers of tomorrow whose names I will never know, please remember one thing: Stephen Ganzberger was my hero.

I owe a debt to my grandfather that I can never repay. I consider myself lucky and privileged to have been able to meet him. Yes, he was my grandfather, but he also just so happened to be the greatest man I've ever met. His longtime friend, Congressman John Dingell, was absolutely correct. Stephen Ganzberger not only stood for all that is good; he *always* took action. But let us remember the millions of others in the Greatest Generation who also took action. Though the world may have only been blessed with their presence for a short time, all of us owe an eternal debt to that ever-dwindling generation of men that did so much while asking for so little in return. They embarked into the jaws of hell unaware of the horrors that awaited them, then buried their trauma deep down upon returning home, hardly to ever speak of their experiences. But despite it all, these children of the Depression, who only knew struggle and hardship, sailed across two vast oceans in the largest armada this world has ever seen. They took action. They fought. Hundreds of thousands died. But in the end, the incredible victory they achieved in the face of overwhelming odds transcended life itself. For when the beaches trembled, it meant that the mere mortal men of the LCIs like my grandfather were taking their stand against tyranny and oppression that will forever echo through the corridors of immortality.

FAMILY APPENDIX

This portion of the book is reserved as a special reference appendix for my family, though I'm sure all will enjoy. It is a collection of miscellaneous letters, documents, records, accounts, and information that relates to Stephen Ganzberger.

Stephen Ganzberger WWII Campaign Summary

Quartermaster 2/c Stephen Ganzberger entitled to wear three battle stars (1 silver; 2 bronze) for participation in the following seven campaigns. Five bronze battle stars equals one silver battle star. Stephen's silver star was earned from his first five campaigns listed below. His sixth and seventh bronze battle stars were earned for the Leyte and Luzon Operations.

1. New Georgia Group Operations
Rendova–Vangunu occupation, July–August 1943 (repelling bombing attack on July 4, 1943).

2. Treasury Islands–Bougainville Operation
Stirling Island landing, November 11, 1943 (Purple Beach 2).

3. Consolidation of Solomon Islands
Green Islands landing, February 20, 1944, and mopping up of northern Solomon Islands, April 28–June 15, 1944.

4. Bismarck Archipelago Operations
Patrol, bombardment, and escort missions in the Northern Solomon Islands, Bismarck Archipelago, and New Guinea Areas, May–September 1944.

5. Western New Guinea Operations
Morotai landing, September 15, 1944 (Red/White Beaches).

6. Leyte Operation
Leyte landing, October 20, 1944 (Red Beach).

7. Luzon Operation
Lingayen Gulf landing, January 4–9, 1945 (White Beach 2 and 3).

Stephen Ganzberger DD Form 215

CERTIFICATE OF RELEASE OR DISCHARGE FROM ACTIVE DUTY		
1. NAME (Last, First, Middle) GANZBERGER, STEPHEN (N)	2. DEPT, COMPONENT AND BRANCH NAVY-USNR	3. SOCIAL SECURITY NUMBER (Also, Service Number if applicable)
4. MAILING ADDRESS (Include ZIP Code) ✕✕✕✕✕✕✕✕✕✕✕		
5. ORIGINAL NAVPERS 553 IS CORRECTED AS INDICATED BELOW:		
ITEM NO.	CORRECTED TO READ	
29.	SEPARATION DATE ON NAVPERS 553 BEING CORRECTED: 8/15/45 SER:62980-09-28542-CR WORLD WAR II VICTORY MEDAL; AMERICAN CAMPAIGN MEDAL; ASIATIC PACIFIC CAMPAIGN MEDAL W/SILVER STAR, COMBAT ACTION RIBBON; PHILIPPINE LIBERATION MEDAL W/2BRONZE STARS; PHILIPPINE PRESIDENTIAL UNIT CITATION (NAVY); AMPHIBIOUS FORCE INSIGNIAS. NO FURTHER ENTRIES	

Stephen Ganzberger List of Military Decorations

Combat Action Ribbon (Navy)
• Awarded for actively participating in ground or surface combat as a member of the United States armed forces.

Philippine Presidential Unit Citation (Philippines)
• This citation badge was awarded by the President of the Republic of the Philippines for extraordinary meritorious service during World War II.

American Campaign Medal
• Awarded for military service in the American Theater of Operations during World War II.

Asiatic–Pacific Campaign Medal (with one silver battle star)
• Awarded for serving in the Asiatic–Pacific Theater in World War II from 1943 to 1945. One silver battle star (3/16 inch) awarded for five bronze battle stars.

Philippine Liberation Medal (with two bronze battle stars)
• Awarded for participation in the Philippines campaign. Two bronze battle stars (3/16 inch) awarded for Leyte and Luzon operations.

World War Two Victory Medal

• Awarded for active-duty naval service between the dates of December 7, 1941 and December 31, 1946.

Amphibious Force Insignia - Shoulder Patch (Navy)

• Awarded for being a member of the U.S. Naval Amphibious Forces, serving aboard a Landing Craft Infantry vessel.

Honorable Service Lapel Button / Honorable Discharge Emblem

• Awarded for having been honorably discharged from the U.S. Navy on August 15, 1945.

Arlington National Cemetery Gravesite

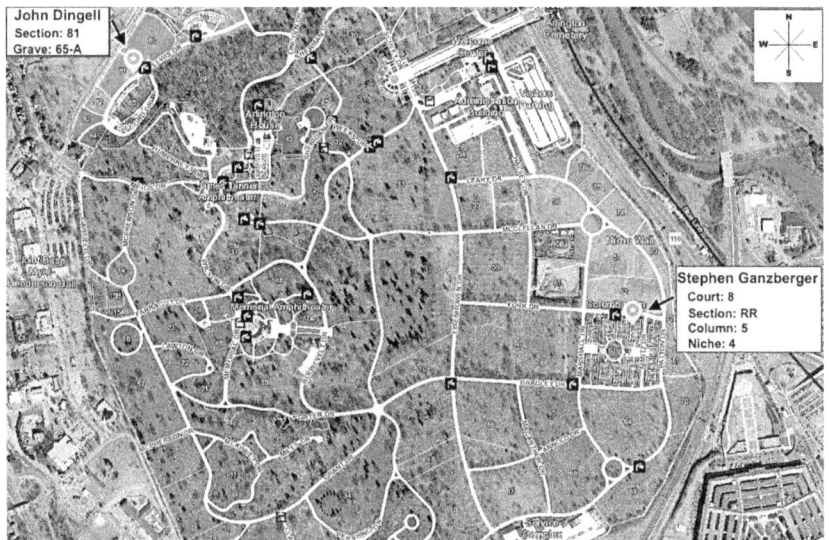

(Official website of U.S. Army)

LETTERS

JOHN D. DINGELL
Fifteenth District
Michigan

May 24, 2011

To my dear friend Patricia Ganzberger, her family and their children:

Each of us can look back on someone who made a great difference in our lives—a person whose wisdom and simple acts of caring made a lasting impression. Steve Ganzberger was such an individual.

From heroic military service to political involvement in the community, he chose to serve others no matter the cost. As Deborah and I reflect on the many years of friendship we have shared, we will always remember a quiet man who sought no recognition for his kindness and capacity to do the right thing.

Steve's pride in his family was evident. You were first in his thoughts and hearts. He enjoyed living and he enjoyed giving.

Today, you honor a man who lived his life with patience and purpose. His children will be able to tell their children that he not only stood for all that is good, he always took action.

With every good wish,

Sincerely yours,

John D. Dingell
Member of Congress

"UNCLE STEVE"

A Nephew's Remembrance

A letter from Frank Shropa to Michael Ganzberger (Stephen's second oldest son) following Stephen's death in 2011

That eventful day that I met your father for the very first time, I remember it distinctly and with such clarity as though the meeting took place just yesterday.

I was standing on the grass in the front yard of your Grandma Vickie's house on Commonwealth Street in Southgate, Michigan (Grandma Vickie owned the house at the time and later sold it to your Mom & Dad). I was facing the front of the house talking to Grandma Vickie when I heard the deep throaty rumbling of hot rod mufflers. I turned around to see a 1951 chartreuse Ford convertible with the top down cruise onto the concrete driveway and stop about five feet away from me. The driver with a wide grin looked down at the astonished, awestruck little boy and said, "Hi Frankie." It was awesome! Until this time I never knew that cars came in paint colors other than black or white, sounded so menacing, or in convertible models. And this guy knew my name! How good is that?

After your Mom got out of the car, she and Grandma Vickie went into the house. Your Dad sat on the top step of the front porch and talked to me. I do not remember any of the conversation, I was too busy absorbing the image before me. You have heard the expression "COOL" before, but this guy was cool before cool came into existence, So Very Cool. He wore a white T-shirt with a pack of Lucky Strike cigarettes rolled up in his left sleeve. Years later, actor James Dean made this a fashion statement-your Dad was years ahead of his time. A closer inspection of this stranger revealed a tattoo on his arm, never having seen a tattoo before, I asked him if it hurt. With his typical Uncle Steve characteristic grin and laugh he said, "no but he would not recommend it for young boys." Your Mom came out of the house and they both got into that magnificent chariot, started the car (it felt as though the ground was convulsing from the power of the engine and the roar of the mufflers), and to my utter dismay waved goodbye as they drove off towards Fort Street.

The next time I saw your Dad was again at Grandma Vickie's house. This time he arrived driving a red and white Harley-Davidson motorcycle with your Mom riding on the back. This was long before the helmet law. Your dad was wearing a motorcycle visored hat, the compulsory T-shirt with cigarettes in the rolled-up sleeve, Levis and motorcycle boots with chains on them. Again, your Dad was way ahead of his time. Years later Marlon Brando would make this attire famous in a motorcycle movie he starred in. Again, you Mom went in the house to talk to Grandma Vickie and Uncle Steve took me for an adventurous ride on the back of that two wheeled motorized contraption and scared the %#@&*+% out of me! I am sure that we only went for a mile or two, but it felt like we went to Arizona and back. When we returned to Grandma Vickie's house and after I got off the Harley, Uncle Steve said, "you can open your eyes now."

Mike, you are probably wondering why I am sharing these experiences with you, I shall attempt to explain. You once told me that I was an inspiration to you. Perhaps life had come full circle from the time when your Dad, the man that I call Uncle Steve, took the time to talk to an insecure, inhibited, immature little boy and treat him as a special important individual. He demonstrated to this kid that there is much more to life than a black or white painted car and better yet why not something absolutely wild such as a motorcycle. It was my first introduction to the concept of why not and I never forgot this lesson. But more than that your Dad exchanged the cool car and the hot Harley for a Willis Jeep, your Mom's everlasting love and the sons and daughters that blessed his life. He worked long and hard while each of you shared a common blood supply with your Mom. Then together, they wove the very fabric of your soul. That is the ultimate Cool!

Stephen Ganzberger is gone. You see God has called him home.
Your Dad is gone. You see God has called him home.

My Uncle Steve is gone. You see God has called him home, and we shall miss him.

But he has not left, he is never gone, he is in your genes and that of your children and their children and their children and for this we are thankful.

Uncle Steve's Nephew,
Frank Shropa

Stephen Ganzberger's son, Michael, shares his favorite memories of his dad:

In 1960–1961, a severe recession hit the United States. Building construction slowed down to a crawl. Dad was laid off and couldn't find work, so he started his own heating/air conditioning service to make ends meet. During the summertime and on weekend service calls, I went along with him as his "gofer" helper (I was six or seven years old at the time), getting tools out of his truck and keeping him company.

On the way home after one service call, we passed a couple selling apples from a card-table off the side of the road. Dad pulled over, chatted with them and bought a paper bag of apples for a couple dollars. A lot of money at the time. When we drove off, I opened the bag and looked at them. They were pitted, bruised and some looked like they had worm holes. Definitely organic! I asked dad why he didn't just get good apples at Kroger when we got home? He said, "if we don't stop to help each other nobody will." I carried that lesson of empathy with me ever since.

Again, in the early Sixties, I remember being with him once when we came up upon a motorcycle accident on a lonely stretch of road in Trenton. It must have just occurred. He pulled over and checked on the guy along with a couple others then he began directing traffic as cars began backing up. Eventually, the police and ambulance arrived. He just wanted to help others. Another lesson. There were more but these were a couple that left early lasting impressions.

Michael Ganzberger
Stephen's second-oldest son

U.S. President Barack Obama
December 2013

The United States of America
honors the memory of

Stephen Ganzberger

This certificate is awarded by a grateful nation in recognition of devoted and selfless consecration to the service of our country in the Armed Forces of the United States.

President of the United States

NOTES

Preface

1. *USN Deck Log, USS LCI (L) 329* – March 16 to December 31, 1943 (450 pages: box 946); Courtesy United States National Archives & Records Administration (NARA), College Park, MD.

   USN Deck Log, USS LCI (G) 65 – December 16, 1942 to August 15, 1945 (1,485 pages: boxes 823–826); Courtesy United States National Archives & Records Administration (NARA), College Park, MD.

Introduction

1. Ernie Pyle, *Brave Men* (Lincoln, NE: University of Nebraska Press/First Bison Books Printing, 2001), p. 17.
2. John Wukovits, *For Crew and Country: The Inspirational True Story of Bravery and Sacrifice aboard the USS Samuel B. Roberts*, (New York, NY: St. Martin's Press, 2013), p. 54.
3. Charles H. Uhl (LCI 362), "Boston to Borneo," *Elsie Item* – Issue 12 (March 1995), p. 7.
4. Robert W. Kirsch, LCI (R) 74, *The Story of a New Ship of War*.
5. Russell W. Hartwell, *Amphibious Assault Landing Craft - USS LCI(L) 711* (Castle Publishers, 1991) p. 11.
6. Harold G. Marquardt, "The Following is a Log and Record of My Experiences While in the Service of the United States Navy," p. 16. *Note*: on March 12, 1944, an Italian refugee gave birth aboard the LCI 15 in the No. 3 hold while underway after departing Anzio. According to Marquardt, this "caused a lot of excitement!!"
7. *Note*: the spelling variations used for full LCI names can differ; other acceptable and correct spellings are LCI(L)-329, LCIL-329, and LCI(L) 329. The author's personal preference in this book includes spaces—LCI (L) 329—as they tend to read easier.
8. Harry Gailey, *Bougainville: The Forgotten Campaign* (Lexington, KY: University Press of Kentucky, 2003).
9. *Note*: some LCIs were converted and renamed as LCI (Demolition), or LCI (D), as well as LCI (Flotilla Flagship), or LCI (FF).

Chapter 1

1. *Note*: Stephen's half brother Bill chose to spell his last name "Gansberger," with an s instead of a z.

2. Gordon Prange, with Donald Goldstein and Katherine Dillon, *December 7 1941: The Day the Japanese Attacked Pearl Harbor* (New York, NY: Wings Books, of Random House Co., 1991).

3. *Note*: a congressional page was hired to serve as a messenger in Congress, often a youth employed as a personal attendant to a person of high rank.

4. John David Dingell Jr., with David Bender, *The Dean: The Best Seat in the House* (New York, NY: HarperCollins Publishers, 2018) pp. 7–18.

5. Ibid, pp. 21–22.

6. John D. Lukacs, *Escape from Davao: The Forgotten Story of the Most Daring Prison Break of the Pacific War* (New York, NY: Simon & Schuster, 2010), pp. 8, 62.

7. John Koster, *Operation Snow: How a Soviet Mole in FDR's White House Triggered Pearl Harbor* (Washington, D.C.: Regnery Publishing, Inc., 2012).

8. John C. Reilly Jr., "Lousy Crate Indeed," *Sea Classics* – Vol. 17, No. 3 (May 1984), p. 62.

Chapter 2

1. Richard W. Johnston (United Press), "Gallantry at Eniwetok," *Elsie Item* – Issue 81 (February 2013), pp. 23–24.

2. Daniel E. Barbey, *MacArthur's Amphibious Navy: Seventh Amphibious Force Operations 1943–1945* (Annapolis, MD: United States Naval Institute, 1969), p. 19. *Note*: Barbey also mentions that original British designs for amphibious assault landing craft had no sleeping accommodations for the troops, however, all American-built LCIs had bunks installed for the troops and ship's crew.

3. *USS LCI Landing Craft Infantry: Vol I* (Paducah, KY: Turner Publishing Co., 1993), p. 8.

4. John H. Morrill, *The Cincinnati* (Wytheville, VA: Wordsprint, Inc., 1994), pp. 1–2.

5. Louis R. Harlan, *All at Sea* (Champaign, IL: University of Illinois Press, 1996), p. 22.

6. John H. Morrill, *The Cincinnati* (Wytheville, VA: Wordsprint, Inc., 1994), pp. 16–17.

7. Chuck Savard, "An LCI at the Battle of Leyte Gulf – and other places!" *Elsie Item* – Issue 50 (October 2004), p. 29.

8. Ernie Pyle, *Brave Men* (Lincoln, NE: University of Nebraska Press/First Bison Books Printing, 2001), p. 12.

9. Source: https://seabeemagazine.navylive.dodlive.mil/files/2014/02/Chapter-3.pdf

10. William L. McGee, *The Amphibians Are Coming! Emergence of the 'Gator Navy and its Revolutionary Landing Craft* (Napa, CA: BMC Publications, 2000), p. 195.

11. *LCI Flotilla Five War Diary*, June 1943.

12. USS LCI(L) 334 Combined War Diary and Log, 24 Nov. 1942 – 30 Nov. 1943. See also: William L. McGee, "*The Amphibians Are Coming! Emergence of the 'Gator Navy and its Revolutionary Landing Craft*" (Napa, CA: BMC Publications, 2000), p. 207.

13. William L. McGee, *The Amphibians Are Coming! Emergence of the 'Gator Navy and its Revolutionary Landing Craft* (Napa, CA: BMC Publications, 2000), p. 209.

14. Oral History, *Reminiscences of John C. Niedermair* (U.S. Naval Institute), p. 240.

15. Melvin D. Barger, *The Genius Behind The LST: John C. Niedermair Was a Giant in*

the Field of Naval Architecture. Source: https://www.uslst.org/memories/27-arti cles/16-lst-memories-the-genius-behind-the-lst

16. Oral History, *Reminiscences of John C. Niedermair* (U.S. Naval Institute), pp. 254–255.

17. Robert W. Kirsch, *The Story of a New Ship of War*.

18. *USS LCI Landing Craft Infantry: Vol I* (Paducah, KY: Turner Publishing Co., 1993), p. 8.

19. Oral History, *Reminiscences of John C. Niedermair* (U.S. Naval Institute), pp. 254–255.

20. Bill D. Ross, *Peleliu Tragic Triumph: The Untold Story of the Pacific War's Forgotten Battle* (New York, NY: Random House, 1991), p. 40.

21. *History of United States Naval Operations in World War II* by Samuel Eliot Morison, copyright © 1966. Reprinted by permission of Little, Brown, & Company, an imprint of Hachette Book Group, Inc.

22. General Douglas MacArthur, *Reminiscences* (New York: McGraw-Hill, 1964), p. 166.

23. An interview with Royal A. Wetzel, conducted by Richard Misenhimer on October 20, 2011 (courtesy National Museum of the Pacific War, Fredericksburg, TX).

24. William L. McGee, *The Amphibians Are Coming! Emergence of the 'Gator Navy and its Revolutionary Landing Craft* (Napa, CA: BMC Publications, 2000), p. 218.

25. Interview with Elmo Pucci LCI-329, conducted by William L. McGee. Transcript now in possession of National Museum of the Pacific War, Fredericksburg, TX.

26. Ibid.

27. Chuck Savard, "An LCI at the Battle of Leyte Gulf – and other places!" *Elsie Item* – Issue 50 (October 2004), p. 29.

28. William H. McCracken, "Can You Top This Sea Story?" *Elsie Item* – Issue 80 (Dec. 2012), p. 24.

29. An interview with Royal A. Wetzel, conducted by Richard Misenhimer on October 20, 2011 (courtesy National Museum of the Pacific War, Fredericksburg, TX).

30. Interview with Elmo Pucci LCI-329, conducted by William L. McGee. Transcript now in possession of National Museum of the Pacific War, Fredericksburg, TX.

31. Ibid.

32. George Weber (LCI 370), "Life Aboard a Flotilla Flagship or It ain't quite exactly what you thought!" *Elsie Item* – Issue 72 (July 2010), p. 11.

33. Ensign Read Dunn Jr., *Remembering*, pp. 22–23.

34. Mitch Weiss, *The Heart of Hell: The Untold Story of Courage and Sacrifice in the Shadow of Iwo Jima* (New York, NY: Berkley Caliber, 2016), pp. 49–50.

35. John Wukovits, *For Crew and Country: The Inspirational True Story of Bravery and Sacrifice aboard the USS Samuel B. Roberts* (New York, NY: St. Martin's Press, 2013), p. 25.

36. Mitch Weiss, *The Heart of Hell: The Untold Story of Courage and Sacrifice in the Shadow of Iwo Jima* (New York, NY: Berkley Caliber, 2016), p. 51.

37. Russell W. Hartwell, *Amphibious Assault Landing Craft - USS LCI(L) 711* (Castle Publishers, 1991), p. 20.

38. Louis R. Harlan, *All at Sea* (Champaign, IL: University of Illinois Press, 1996), p. 23.

39. Oral History, *Reminiscences of John C. Niedermair* (U.S. Naval Institute), p. 256.

40. Ernie Pyle, *Brave Men* (Lincoln, NE: University of Nebraska Press/First Bison Books Printing, 2001), p. 96.

41. Louis R. Harlan, *All at Sea* (Champaign, IL: University of Illinois Press, 1996), pp. 30, 175.

42. John H. Morrill, *The Cincinnati* (Wytheville, VA: Wordsprint, Inc., 1994), p. 16.

43. "Vice Admiral Lorenzo Sherwood Sabin Jr., May 23, 1899 – June 2, 1988," *Elsie Item* – Issue 93 (April 2016), pp. 12–13.

44. Louis R. Harlan, *All at Sea* (Champaign, IL: University of Illinois Press, 1996), pp. 31–42.

45. John Hersey, "U.S.S. LCI 226: After her Pacific Crossing and for actions this ugly craft becomes a real fighting ship," *Life*, March 27, 1944, p. 53.

46. Bob L. Petit, "Bob Petit's Story: The Trials and Triumphs of an Underage LCI Sailor!" *Elsie Item* – Issue 64 (August 2008), p. 19.

47. Mitch Weiss, *The Heart of Hell: The Untold Story of Courage and Sacrifice in the Shadow of Iwo Jima* (New York, NY: Berkley Caliber, 2016), p. 28.

48. Robert W. Kirsch, *The Story of a New Ship of War.*

49. Louis R. Harlan, *All at Sea* (Champaign, IL: University of Illinois Press, 1996), p. 32.

50. Thomas E. Woodstrup, "LCI (FF) 628," *Elsie Item* – Issue 15 (December 1995), p. 13.

51. Bill "Big Bill" Athan, "The Saga of the LCI 446: Salty Tales of the South Pacific!" *Elsie Item* – Issue 43 (January 2003), p. 23.

52. "Vice Admiral Lorenzo Sherwood Sabin Jr. May 23, 1899 – June 2, 1988," *Elsie Item* – Issue 93 (April 2016), p. 13.

53. Robert V. Rosenwald, "USS LCI(L) 1008," *Elsie Item* – Issue 17 (June 1996), p. 13.

54. John Miller Jr., *Cartwheel: The Reduction of Rabaul* (Washington, D.C.: Office of the Chief of Military History Department of the Army, 1959), pp. vii, 25.

55. General Douglas MacArthur, *Reminiscences* (New York: McGraw-Hill, 1964), p. 165.

56. Louis R. Harlan, *All at Sea* (Champaign, IL: University of Illinois Press, 1996), p. 31.

Chapter 3

1. Gordon L. Rottman, *Landing Craft, Infantry and Fire Support* (Long Island, NY: Osprey Publishing Ltd., 2009), p. 43.

2. William Doyle, *PT-109: An American Epic of War, Survival, and the Destiny of John F. Kennedy* (HarperCollins Publishers, 2015) p. 46. See also: "KILL JAPS": William Manchester, *The Glory and the Dream: A Narrative History of the United States, 1932–1972* (Little, Brown, 1974), p. 329.

3. William L. McGee, *The Solomons Campaigns 1942–1943 From Guadalcanal to Bougainville Pacific War Turning Point* (Napa, CA: BMC Publications, 2001), p. 347.

4. USS LCI(L) 329 War Diary, June 23–July 31, 1943, p. 3.

5. Howard G. Sawyer, "This is the Story of the LCI 336," *Elsie Item* – Issue 27 (December 1998), p. 24.

6. Eric Hammel, *Munda Trail: The New Georgia Campaign* (New York, NY: Orion Books/Crown Publishers, 1989), p. 77. *Note*: Lou Plant and Lt. Col. Henry Shafer

identified the sixteen Japanese bombers as "Bettys" (Mitsubishi land based G4M Bombers). However, the official Action Reports from July 4, 1943, from the LCIs, and several LSTs, state that the sixteen bombers were identified as "Sallys" (Mitsubishi Ki-21 Type 97 Heavy Bombers). It is entirely possible that there was a combination of both types of bombers. It should also be noted that Lou Plant remembers only one of the sixteen Japanese bombers being destroyed.

7. LCI (L) 329 War Diary, June–July 1943. See also: Official Action Report: USS LCI (L) 24, July 4, 1943 (pp. 9–10); Action Reports: Commander, South Pacific, dated October 17, 1943: Anti-Aircraft Action During Landings at Rendova and Vella Lavella (25 pages)

8. Melville Bell Grosvenor, *The National Geographic Magazine* Vol. LXXXVI, No.1 (Washington, July 1944), p. 22.

9. Louis V. Plant, *Memories of World War II*, p. 17.

10. Ibid.

11. *USS LCI Landing Craft Infantry: Vol I* (Paducah, KY: Turner Publishing Co., 1993), p. 2.

12. Louis V. Plant, *Memories of World War II*, p. 18.

13. Ernie Pyle, *Brave Men* (Lincoln, NE: University of Nebraska Press/First Bison Books Printing, 2001), p. 452.

14. Howard G. Sawyer, "This is the Story of the LCI 336," *Elsie Item* – Issue 27 (December 1998), p. 24.

15. *Official Action Report: USS LCI (L) 24; July 4, 1943*, pp. 9–10.

16. *War Diary: USS LCI (L) 329*, July 1943, pp. 5–6.

17. *Note*: the correspondents were Edmundson of *Fortune* and Boddie of *Newsweek*.

18. An interview with Royal A. Wetzel, conducted by Richard Misenhimer on October 20, 2011 (courtesy National Museum of the Pacific War, Fredericksburg, TX).

19. Eric Hammel, *Munda Trail: The New Georgia Campaign* (New York, NY: Orion Books/Crown Publishers, 1989), pp. 95, 182, 218.

Chapter 4

1. John H. Morrill, *The Cincinnati* (Wytheville, VA: Wordsprint, Inc., 1994), p. 29. *Note*: This message was composed by Bob Neff and J.P. "Red" Smith and sent to all hands of LCI "Black Cat" Flotilla Thirteen in December 1943.

2. Ernie Pyle, *Brave Men* (Lincoln, NE: University of Nebraska Press/First Bison Books Printing, 2001), pp. 36–37.

3. *Note*: the first case had been one of the 329's coxswains named J. Gilbert who was transferred to a hospital on Florida Island on July 26, 1943.

Chapter 5

1. William L. McGee, *The Amphibians Are Coming! Emergence of the 'Gator Navy and its Revolutionary Landing Craft* (Napa, CA: BMC Publications, 2000), p. 246.

2. *USS LCI Landing Craft Infantry: Vol II* (Paducah, KY: Turner Publishing Co., 1995), p. 67.

3. *War Diary: USS LCI (L) 329*, October 1943, p. 2.

4. Robin L. Rielly, *American Amphibious Gunboats in World War II: A History of the LCI and LCS(L) Ships in the Pacific* (Jefferson, NC: McFarland & Company Publishers, 2013), pp. 78–79.

5. Charles R. Ports, "My Life as an LCI Gunboat Sailor," *Elsie Item* – Issue 56 (June 2006), p. 20.

6. USS LCI(L) 23, *Action Report: Support of Landing, Treasury Islands, 27 October 1943*, pp. 29–30.

7. Ibid, p. 33B.

8. Ibid, p. 33C.

9. *Note*: amtracs, or Landing Vehicle Tracked (LVTs) were specially designed personnel landing crafts with tracks on them. Also referred to as "amtraks" or "amphtracs" throughout this book.

10. John Wukovits, *One Square Mile of Hell: The Battle for Tarawa* (New York, NY: NAL Caliber/Penguin Group, 2006), p. 112. See also: Ota and Williams, "Tarawa, My Last Battle," p. 5.

11. Charles P. Grow, typewritten personal account, "Reminiscences of the LCI 345," dated January 1983.

12. The first four LCIs converted to gunboats used at Treasury Islands were LCIs 22, 23, 21, and 70.

13. William L. McGee, *The Amphibians Are Coming! Emergence of the 'Gator Navy and its Revolutionary Landing Craft* (Napa, CA: BMC Publications, 2000), pp. 204–205.

Chapter 6

1. Charles R. Ports, "My Life as an LCI Gunboat Sailor," *Elsie Item* – Issue 56 (June 2006), p. 20.

2. John Hersey, "U.S.S. LCI 226: After her Pacific Crossing and for actions this ugly craft becomes a real fighting ship," *Life*, March 27, 1944, p. 57.

3. Ibid, p. 61.

4. Ibid.

Chapter 8

1. Douglas MacArthur, *Reminiscences* (New York: McGraw-Hill, 1964), p. 187.

2. Source: https://www.history.navy.mil/content/history/nhhc/research/library/online-reading-room/title-list-alphabetically/b/building-the-navys-bases/building-the-navys-bases-vol-2.html#1-25

3. Mitch Weiss, *The Heart of Hell: The Untold Story of Courage and Sacrifice in the Shadow of Iwo Jima* (New York, NY: Berkley Caliber, 2016), p. 27.

4. Louis R. Harlan, *All at Sea* (Champaign, IL: University of Illinois Press, 1996), p. 26.

5. Ibid, p. 161.

Chapter 9

1. As previously noted in the Introduction, this wasn't always the case. A small number of LCIs were converted and reclassified as LCI (Demolition), or LCI (D), as well as LCI (Flotilla Flagship), or LCI (FF).
2. Robert W. Kirsch, *The Story of a New Ship of War*. *Note:* "hoppers" was another name for toilets.
3. Mitch Weiss, *The Heart of Hell: The Untold Story of Courage and Sacrifice in the Shadow of Iwo Jima* (New York, NY: Berkley Caliber, 2016), p. 93.
4. *Note:* J.R. Reid remembers there being only four .50 Caliber machine guns on the LCI (G) 65, even though there were six of them according to official records. It is possible that out of six originally installed, two may have been removed.
5. James Douglass Robertson, *Robby* (2012), p. 182.
6. Gilbert Ortiz, in email to Joseph Ortiz (Subj: Re: Joey from Texas), dated September 15, 2008, 6:45 p.m.
7. LeRoy Winston, "An LCI Creed," *Elsie Item* – Issue 12 (March 1995), p. 4.

Chapter 10

1. *Note:* "Washing Machine Charlie" was a name originally given by the Allies (primarily the United States) to Imperial Japanese aircraft that performed flyovers over Henderson Field during the early days of the Guadalcanal campaign. The term, which came from the distinctive sound of the beat-up Japanese aircraft engine, was thereafter broadly used to describe Japanese flyovers at night.

Chapter 11

1. Daniel E. Barbey, *MacArthur's Amphibious Navy: Seventh Amphibious Force Operations 1943–1945* (Annapolis, MD: United States Naval Institute, 1969), p. 190.
2. Admiral William F. Halsey, letter to LST Flotilla Five and SoPac., dated May 5, 1944.

Chapter 12

1. An interview with Royal A. Wetzel, conducted by Richard Misenhimer on October 20, 2011 (courtesy National Museum of the Pacific War, Fredericksburg, TX).
2. Ernie Pyle, edited with a biographical essay by David Nichols, *Ernie's War: The Best of Ernie Pyle's World War II Dispatches* (New York, NY: Touchstone/Simon & Schuster, Inc., 1986), p. 370. Reprinted by permission of E.W. Scripps Company.
3. *USS LCI Landing Craft Infantry: Volume II* (Paducah, KY: Turner Publishing, 1995), p. 64.
4. Charles R. Ports, "My Life as an LCI Gunboat Sailor," *Elsie Item* – Issue 56 (June 2006), p. 21.
5. An interview with Royal A. Wetzel, conducted by Richard Misenhimer on October, 20, 2011 (courtesy National Museum of the Pacific War, Fredericksburg, TX).

6. Ibid.

7. *USS LCI Landing Craft Infantry: Vol I* (Paducah, KY: Turner Publishing Co., 1993), p. 27.

Chapter 13

1. Ernie Pyle, "The God-Damned Infantry," In the Front Lines Before Mateur, Northern Tunisia, dated May 2, 1943. Reprinted by permission of E.W. Scripps Company.

Chapter 14

1. John P. Cummer, "Who We Were . . . Who We Are: Some Memorial Day Thoughts from the Editor," Elsie Item – Issue 41 (June 2002), p. 4

2. John Wukovits, *For Crew and Country: The Inspirational True Story of Bravery and Sacrifice Aboard the USS Samuel B. Roberts*, (New York, NY: St. Martin's Press, 2013), pp. 89–90.

3. William E. Keeler, "Report of LCI Duty from 1944 to 1947."

4. Daniel E. Barbey, *MacArthur's Amphibious Navy: Seventh Amphibious Force Operations 1943–1945* (Annapolis, MD: United States Naval Institute, 1969), p. 230.

Chapter 15

1. Ernie Pyle, edited with a biographical essay by David Nichols, *Ernie's War: The Best of Ernie Pyle's World War II Dispatches* (New York, NY: Touchstone/Simon & Schuster, Inc., 1986), p. 364. Reprinted by permission of E.W. Scripps Company.

2. *History of United States Naval Operations in World War II* by Samuel Eliot Morison, copyright © 1966. Reprinted by permission of Little, Brown, & Company, an imprint of Hachette Book Group, Inc.

3. Daniel E. Barbey, *MacArthur's Amphibious Navy: Seventh Amphibious Force Operations 1943–1945* (Annapolis, MD: United States Naval Institute, 1969), p. 224.

4. Chester Nez with Judith Schiess Avila, *Code Talker* (New York, NY: Penguin Group, 2011), p. 206.

5. *History of United States Naval Operations in World War II* by Samuel Eliot Morison, copyright © 1966. Reprinted by permission of Little, Brown, & Company, an imprint of Hachette Book Group, Inc.

6. Official Action Report, U.S.S. LCI (G) 65, Morotai Operation, dated September 18, 1944, p. 5.

7. Ibid.

8. Ken Howdeshell, in email to the Author (No Subject), dated December 5, 2012, 3:36 p.m.

9. Ken Howdeshell, in email to the Author (Subject: Summary: LCI (G) 65 - Action Report/War Diary Morotai Island Operation), dated December 26, 2021, 1:21 p.m.

10. Official Action Report, U.S.S. LCI (G) 65, Morotai Operation, dated September 18, 1944, pp. 2–6.

11. *History of United States Naval Operations in World War II, Volume XII:* Leyte (June 1944–January 1945), by Samuel Eliot Morison, p. 25.

12. Thomas J. Cutler, *The Battle of Leyte Gulf: 23–26 October 1944*, (New York, NY: Harper Collins Publishers, 1994), pp. 265–266. See also: Kathleen Thomas, *Don't Call Me Rosie: The Women who Welded the LSTs and the Men Who Sailed on Them* (Tigard, OR: Thomas/Wright, Inc., 2004), p. 72. *Note:* the Japanese military term for the kamikazes was *tokkōtai,* meaning "special attack force."

Chapter 16

1. Dan King, *The Last Zero Fighter: Firsthand Accounts from WWII Japanese Naval Pilots* (North Charleston, SC: Pacific Press, 2012), p. 224.

2. *National Geographic:* https://www.nationalgeographic.com/travel/destinations/asia/philippines/partner-content-know-before-you-go-the-philippines/

3. Daniel E. Barbey, *MacArthur's Amphibious Navy: Seventh Amphibious Force Operations 1943–1945* (Annapolis, MD: United States Naval Institute, 1969), pp. 227–229.

4. A. M. Holmes, Commander, USS LCI (G) Group FORTY-FIVE, *War Diary October 1944 for LCI (G) Group FORTY-FIVE,* p. 1.

5. Daniel E. Barbey, *MacArthur's Amphibious Navy: Seventh Amphibious Force Operations 1943–1945* (Annapolis, MD: United States Naval Institute, 1969), p. 229.

6. A. M. Holmes, Commander, USS LCI (G) Group FORTY-FIVE, *War Diary October 1944 for LCI (G) Group FORTY-FIVE,* p. 2.

7. John David Dingell Jr., with David Bender, *The Dean: The Best Seat in the House* (New York, NY: HarperCollins Publishers, 2018), pp. 89–90.

8. Ibid, pp. 91–92.

9. Ibid, p. 92.

10. Commanding Officer, USS LCI (G) 331, *Action report-CENTRAL PHILIPPINES Operation,* November 5, 1944, p. 3. *Note:* USS LCI (G)s 68 and 70, part of LCI (G) Group Forty-Five were assigned to the Southern Attack Force at Green Beach. USS LCI (G)s 64 and 69 were assigned to White Beach. See also: A. M. Holmes, Commander, USS LCI (G) Group FORTY-FIVE, *War Diary October 1944 for LCI (G) Group FORTY-FIVE,* p. 2.

11. *Action Report: Commander Task Unit 78* – Operation Plan No. 101-44, p. 173.

12. Commander Task Force 78 (Commander Seventh Amphibious Force), Serial 00911, *Leyte Operations – Report On,* November 10, 1944, p. 13.

13. Commanding Officer, USS LCI (G) 71, *Action Report, Central Philippine Islands,* November 7, 1944, p. 2.

14. Robin L. Rielly, *American Amphibious Gunboats in World War II: A History of the LCI and LCS(L) Ships in the Pacific* (Jefferson, NC: McFarland & Company Publishers, 2013), p. 149.

15. Daniel E. Barbey, *MacArthur's Amphibious Navy: Seventh Amphibious Force Operations 1943–1945* (Annapolis, MD: United States Naval Institute, 1969), p. 245.

16. Ibid, p. 263.

17. This speech was broadcast over the "Voice of Freedom" circuit. See also: Douglas MacArthur, *Reminiscences,* pp. 142–145, 215–218; and "Battle for the Philippines," *Fortune* XXXI No. 6 (June 1945) pp. 157–158.

18. *Note*: In military terminology, the subsequent days after an invasion were referred to as the letter designation of the invasion followed by a plus sign and the number of days that have gone by since invasion. For example, the first day after the A-Day invasion on October 20 was known as A-plus one day, or A+1. The same logic applies to the day prior to invasion, known as A-minus one day, or A–1.

19. Robert Nichols, *The First Kamikaze Attack?* (May 20, 2020): https://www.awm.gov.au/wartime/28/kamikaze-attack

20. Trevor Jensen, "Walter Simmons, 1908–2006: Editor and War Reporter," *Chicago Tribune*, December 1, 2006. See also: Mitchell Zuckoff, *Lost in Shangri-La: A True Story of Survival, Adventure, and the Most Incredible Rescue Mission of World War II* (New York, NY: HarperCollins, 2011), pp. 220–221.

21. Russell W. Hartwell, *Amphibious Assault Landing Craft - USS LCI(L) 711* (Castle Publishers, 1991), p. 55.

22. Elmer Kinsinger, "Letters to the Editor," *Elsie Item* – Issue 53 (April 2005), p. 6.

23. Ibid.

24. Author's interview with J.R. Reid, February 5, 2014.

25. Elmer Kinsinger, "Letters to the Editor," *Elsie Item* – Issue 53 (April 2005), p. 6.

26. Commanding Officer, U.S.S. LCI (G) 65, *Action Report, Initial Landing Leyte Island, Philippine Islands* (13 November 1944), pp. 1–4.

27. Daniel E. Barbey, *MacArthur's Amphibious Navy: Seventh Amphibious Force Operations 1943–1945* (Annapolis, MD: United States Naval Institute, 1969), p. 262.

28. USS LST 464 War Diary, *Rep of ops in the invasion of Leyte Is, Philippines on 10/24/44–12/15/44*, p. 5.

29. Elmer Kinsinger, "Letters to the Editor," *Elsie Item* – Issue 53 (April 2005), p. 6.

30. The first man was Hurley E. Christian who had been killed in action at the bombing of Rendova Island on July 4, 1943.

31. Commander L. R. McDowell, Commander USS *Blue Ridge* (AGC2), *War Diary*, dated November 13, 1944, p. 4.

32. Daniel E. Barbey, *MacArthur's Amphibious Navy: Seventh Amphibious Force Operations 1943–1945* (Annapolis, MD: United States Naval Institute, 1969), p. 262.

33. *USS LCI Landing Craft Infantry: Volume II* (Paducah, KY: Turner Publishing, 1995), p. 80.

34. *Action Report: Commander Task Unit 78* – Operation Plan No. 101-44, p. 151.

35. C. Vann Woodward, *The Battle For Leyte Gulf: The Incredible Story of World War II's Largest Naval Battle* (New York, NY: Skyhorse Publishing, Inc., 2007), pp. 30–31.

36. *Note*: The six battleships present were: USS *Pennsylvania, California, Tennessee, Mississippi, Maryland*, and *West Virginia*.

37. Samuel Eliot Morison, *History of United States Naval Operations in World War II, Volume XII: Leyte* (June 1944–January 1945), p. 240.

38. James D. Hornfischer, *The Fleet at Flood Tide: America at Total War in the Pacific, 1944–1945* (New York, NY: Bantam Books, 2016), p. 356.

39. C. Vann Woodward, *The Battle For Leyte Gulf: The Incredible Story of World War II's Largest Naval Battle* (New York, NY: Skyhorse Publishing, Inc., 2007), pp. 75–76.

40. John Wukovits, *For Crew and Country: The Inspirational True Story of Bravery and*

Sacrifice aboard the USS Samuel B. Roberts, (New York, NY: St. Martin's Press, 2013), p. 120.

41. *Note*: The LCI (G) 23 suffered much damage and casualties from two near misses by what was believed to be 250-pound bombs. Officially, five enlisted men were killed outright, four died later, two officers and twenty-seven men were injured, although one account claims ten men were KIA. The LCI 23 was badly holed on the starboard side and rendered inoperative. See: A. M. Holmes, Commander, USS LCI (G) Group FORTY-FIVE, *War Diary October 1944 for LCI (G) Group FORTY-FIVE*, p. 4.

42. Commanding Officer, USS LCI (G) 338, *Action Report – Central Philippines Operation*, p. 4.

43. Charles R. Ports, "My Life as an LCI Gunboat Sailor," *Elsie Item – Issue 56* (June 2006), p. 22.

44. Rpt, Capt Ray Tarbuck, USN, Observers Rpt of KING II Opn, 3 Nov 44, GHQ G-3 Jnl, 30 Oct 44.

45. U.S.S. LST 552 *Action Report*, dated October 31, 1944, pp. 1–13.

46. *Note*: Rear Admiral Clifton Sprague, whose nickname was "Ziggy" (not to be confused with overall commander of Escort Carrier Task Group 77.4, Rear Admiral Thomas L. Sprague), was the commander of task unit Taffy 3 (TG 77.4.3).

47. *History of United States Naval Operations in World War II* by Samuel Eliot Morison, copyright © 1966. Reprinted by permission of Little, Brown, & Company, an imprint of Hachette Book Group, Inc.

48. *Note*: According to Barbey, LCIs 34, 71, 337, 340, and 341 were involved in rescuing Taffy 3 survivors. The four ships of Taffy 3 sunk from Kurita's Central Force on Oct. 25 were *Samuel B. Roberts*, *Hoel*, *Johnston*, and *Gambier Bay*. The *St Lo* was sunk later that day by a kamikaze.

49. John Wukovits, *For Crew and Country: The Inspirational True Story of Bravery and Sacrifice aboard the USS Samuel B. Roberts*, (New York, NY: St. Martin's Press, 2013), p. 225. Originally from Tom Stevenson interview, March 27, 2001; "Samuel B. Roberts Hit Jap Cruiser in Torpedo Attack," p. 3.

50. Dominick Maurone, "Voices from the Pacific," *Elsie Item – Issue 71* (May 2010), p. 27.

51. Russell W. Hartwell, *Amphibious Assault Landing Craft - USS LCI(L) 711* (Castle Publishers), p. 55.

52. A. M. Holmes, Commander, USS LCI (G) Group FORTY-FIVE, *War Diary October 1944 for LCI (G) Group FORTY-FIVE*, p. 4.

53. Commanding Officer, U.S.S. *Quapaw* (AT-110), *Action Reports – Philippines Operation – For Period 29 October to 14 November 1944*, p. 5.

Chapter 17

1. Russell W. Hartwell, *Amphibious Assault Landing Craft - USS LCI(L) 711* (Castle Publishers), p. 59.

2. *Action Report,* U.S.S. *Achilles* (ARL-41), dated November 16, 1944, p. 1.

3. USS LCI (L) 432, *Action Report – Anti-Aircraft, Leyte Operation, narrative form*, dated November 12, 1944, p. 4.

4. Ibid.
5. Ibid.
6. Ibid, pp. 4–5.
7. USS LCI (G) 65, *Deck Log*, Sunday, November 12, 1944.
8. USS *Achilles* (ARL 41), *Action Report*, dated November 16, 1944, pp. 3–4. KIA figures include fifteen men listed as MIA.
9. Charles J. Macaluso, USS LCI (G) 65, *Action Report – Anti-Aircraft by Surface Ships*, November 12, 1944, pp. 3–4.
10. Commanding Officer, USS LST 464, *Action Report – Leyte Gulf Operation*. (October 24–December 15, 1944), p. 9.
11. Daniel E. Barbey, *MacArthur's Amphibious Navy: Seventh Amphibious Force Operations 1943–1945* (Annapolis, MD: United States Naval Institute, 1969), p. 278.
12. Robert J. Cressman, *The Official Chronology of the U.S. Navy in World War II* (Annapolis, MD: Naval Institute Press, 2000), pp. 273–274.
13. *Action Report – USS LST 464* (October–December, 1944), p. 10.

Chapter 18

1. Walter Krueger, *From Down Under to Nippon: The Story of the Sixth Army in World War II* (1963), p. 187. See also: Nathan N. Prefer, *Leyte 1944: The Soldiers' Battle* (Havertown, PA: Casemate Publishers, 2012).
2. *Action Report – USS LST 464* (October–December, 1944), p. 10.
3. Author's interview with Oliver J. Budd on August 24, 2014.
4. *The Notre Dame Alumnus* – October 1936 issue, Archives of the University of Notre Dame, p. 13.
     http://www.archives.nd.edu/Alumnus/VOL_0015/VOL_0015_ISSUE_0001.pdf

Chapter 19

1. O.N.I. Weekly III 3998 (December 13, 1944).
2. Sixth Army Report III 27. See also: *History of United States Naval Operations in World War II, Volume XIII: The Liberation of the Philippines Luzon, Mindanao, the Visayas 1944–1945*, by Samuel Eliot Morison, pp. 3–13, 111.
3. Gilbert Ortiz, in email to Joseph Ortiz (Subj: Re: Joey from Texas), dated September 15, 2008, 6:45 p.m.
4. W. J. McKeon, *U.S.S. LCI (G) 65, Action Report, Lingayen Operation*, pp. 1–6.
5. Richard Call, Commanding officer, USS LCI (G) 64 *Action Report – Luzon Operations* (dated January 15, 1945), pp. 1–11.
6. James Douglass Robertson, *Robby* (2012), p. 226.
7. Ibid, p. 227.
8. Ibid, p. 228.
9. Ibid, p. 226.
10. An interview with Royal A. Wetzel, conducted by Richard Misenhimer on October 20, 2011 (courtesy National Museum of the Pacific War, Fredericksburg, TX).
11. An interview with John A. Reulet, conducted by Richard Misenhimer on

October 4, 2011 (courtesy National Museum of the Pacific War, Fredericksburg, TX).

12. An interview with Leo Wilcox, conducted by Richard Misenhimer on December 1, 2011 (courtesy National Museum of the Pacific War, Fredericksburg, TX).

13. Gilbert Ortiz, in email to Joseph Ortiz (Subj: Re: Joey from Texas), dated September 15, 2008, 6:45 p.m.

14. An interview with John A. Reulet, conducted by Richard Misenhimer on October 4, 2011 (courtesy National Museum of the Pacific War, Fredericksburg, TX).

15. Gilbert Ortiz, in email to Joseph Ortiz (Subj: Re: Joey from Texas), dated September 15, 2008, 6:45 p.m.

16. *History of United States Naval Operations in World War II by Samuel Eliot Morison,* copyright © 1966. Reprinted by permission of Little, Brown, & Company, an imprint of Hachette Book Group, Inc.

17. Daniel E. Barbey, *MacArthur's Amphibious Navy: Seventh Amphibious Force Operations 1943–1945* (Annapolis, MD: United States Naval Institute, 1969), pp. 296–297.

18. *History of United States Naval Operations in World War II* by Samuel Eliot Morison, copyright © 1966. Reprinted by permission of Little, Brown, & Company, an imprint of Hachette Book Group, Inc. See also: *The Divine Wind,* pp. 219–220; *Note:* MacArthur Historical Report II 432 says that 58 kamikazes and 17 escorts were involved.

19. Dick Camp, *Iwo Jima Recon: The U.S. Navy at War, February 17, 1945* (St. Paul, MN: Zenith Press, 2007), pp. 48–49.

20. *Note:* As mentioned, there are three conditions of Battle Stations:

Condition I is General Quarters. Under Condition I, all battle stations are manned, and usually surface or air action is imminent (about to take place). Condition I is sometimes modified to let a few persons at a time rest on station or to let designated personnel draw rations for delivery to battle stations (condition IE).

Condition II is a special watch used by gunfire support ships for situations such as extended periods of shore bombardment.

Condition III is the normal wartime cruising watch. Normally, when cruising under Condition III, the ship's company stands watch on a basis of four hours on, eight hours off; about one-third of the ship's armament is manned in the event of a surprise attack.

Source:

https://seabeemagazine.navylive.dodlive.mil/files/2014/02/Chapter-3.pdf

21. James Douglass Robertson, *Robby* (2012), pp. 242–244.

Chapter 20

1. USS LCI (G) 473, *Action Report on Invasion of Iwo Jima,* dated August 13, 1945. *Note:* this endorsement by Admiral Nimitz was one of the finest recognitions of the LCIs in World War II.

2. Robert Harker, "The LCI Gunboats at Iwo Jima," *Elsie Item* – Issue 79 (June 2012), p. 17.

3. Garnett A. Ridenhour, "The War Years," *Elsie Item* – Issue 32 (March 2000), pp. 21–23.

Chapter 21

1. E.B. Sledge, *With the Old Breed at Peleliu and Okinawa* (New York: Ballantine Books, 1981), p. 315.
2. W. Donald Stewart, "LCI(R) 706," *Elsie Item* – Issue 21 (June 1997), p. 12.
3. E.B. Sledge, *With the Old Breed at Peleliu and Okinawa* (New York: Ballantine Books, 1981), p. 253.
4. John David Dingell Jr., with David Bender, *The Dean: The Best Seat in the House* (New York, NY: HarperCollins Publishers, 2018) p. 94.
5. Ibid, p. 95.
6. Joe L. Harris, *My Adventures in World War II*, p. 6.
7. An interview with Royal A. Wetzel, conducted by Richard Misenhimer on October 20, 2011 (courtesy National Museum of the Pacific War, Fredericksburg, TX).
8. Harold R. Ronson, "Remembering Franklin Delano Roosevelt," *Elsie Item* – Issue 32 (March 2000), p. 34.
9. Report of Medical Survey (NMS—Form M, 1940), Ganzberger, Stephen (U.S. Naval Hospital: Great Lakes, Illinois; dated July 30, 1945).
10. John David Dingell Jr., with David Bender, *The Dean: The Best Seat in the House* (New York, NY: HarperCollins Publishers, 2018), p. 96.
11. Robert V. Rosenwald, "USS LCI(L) 1008," *Elsie Item* – Issue 17 (June 1996), p. 13.
12. LCI 65 info: https://www.navsource.org/archives/10/15/150065.htm
    LCI 329 info: https://www.navsource.org/archives/10/15/150329.htm
13. Robert W. Kirsch, *The Story of a New Ship of War.*
14. http://usslci.org/facts/

Chapter 22

1. Daniel E. Barbey, *MacArthur's Amphibious Navy: Seventh Amphibious Force Operations 1943–1945* (Annapolis, MD: United States Naval Institute, 1969), p. 30.
2. Letter to John Dingell from E. A. Wright, Department of the Navy, dated September 3, 2009.
3. John H. Morrill, *The Cincinnati* (Wytheville, VA: Wordsprint, Inc., 1994), pp. 15, 67. *Note*: Lucius Quinctius Cincinnatus (c. 519–c. 430 B.C.) was a famous Roman soldier and statesman who, after faithful military service, laid down all of his powerful offices and returned to civilian life.
4. James Douglass Robertson, *Robby* (2012), pp. 182–183.
5. Dr. R. William Clark, *Elsie Item* – Issue 14 (September 1995), p. 12.
6. Patricia Rone, "Letters to the Editor," *Elsie Item* – Issue 61 (October 2007), p. 9.
7. *Note*: congressional redistricting caused the Dingells to later represent Michigan's 12th Congressional district instead of the 15th they'd previously been representing.

Epilogue

1. Richard W. Johnston (United Press), "Gallantry at Eniwetok," *Elsie Item* – Issue 81 (February 2013), pp. 23–24.

BIBLIOGRAPHY

Barbey, Daniel E. *MacArthur's Amphibious Navy: Seventh Amphibious Force Operations 1943-1945*. Annapolis: United States Naval Institute, 1969.

"Battle for the Philippines." *Fortune XXXI No. 6* (1945): 157-158.

Camp, Dick. *Iwo Jima Recon: The U.S. Navy at War, February 17, 1945*. St. Paul: Zenith Press, 2007.

Cannon, M.H. *Leyte: The Return to the Philippines, U.S. Army in WWII, The War in the Pacific*. 1993.

Cressman, Robert J. *The Official Chronology of the U.S. Navy in World War II*. Annapolis: Naval Institute Press, 2000.

Cutler, Thomas J. *The Battle of Leyte Gulf: 23-26 October 1944*. New York: Harper Collins Publishers, 1994.

Dingell Jr., John David and David Bender. *The Dean: The Best Seat in the House*. New York: HarperCollins Publishers, 2018.

Doyle, William. *PT-109: An American Epic of War, Survival, and the Destiny of John F. Kennedy*. New York: HarperCollins Publishers, 2015.

Dunn Jr., Read. *Remembering*. n.d.

Finkler, Jim and William D. Elder. "Rough Ride Across the Atlantic: A Coast Guardsman's Story." *Elsie Item 77* (2011): 6-9.

Gailey, Harry. *Bougainville: The Forgotten Campaign*. Lexington: University Press of Kentucky, 2003.

Grosvenor, Melville Bell. *The National Geographic Magazine LXXXVI, No.1*. Washington, 1944.

Hammel, Eric. *Munda Trail: The New Georgia Campaign*. New York: Orion Books/Crown Publishers, 1989.

Harker, Robert. "The LCI Gunboats at Iwo Jima." *Elsie Item* 79 (2012): 17-9.

Harlan, Louis R. *All at Sea*. Champaign: University of Illinois Press, 1996.

Harris, Joe L. *My Adventures in World War II* . n.d.

Hartwell, Russell Worden. *Amphibious Assault Landing Craft - USS LCI(L) 711*. Spanaway, WA: Castle Publishers, 1991.

Hersey, John. "U.S.S. LCI 226: After her Pacific Crossing and for actions this ugly craft becomes a real fighting ship." *Life* 27 March 1944: 53–61.

Hillenbrand, Laura. *Unbroken: A World War II Story of Survival, Resilience, and Redemption*. New York: Random House, Inc., 2010.

Hornfischer, James D. *The Fleet at Flood Tide: America at Total War in the Pacific, 1944-1945*. New York: Bantam Books, 2016.

Jensen, Trevor. "Walter Simmons, 1908–2006: Editor and War Reporter," *Chicago Tribune*, 2006.

Johnston, Richard W. "Gallantry at Eniwetok." *Elsie Item* 81 (2013): 23-24.

King, Dan. *The Last Zero Fighter: Firsthand Accounts from WWII Japanese Naval Pilots*. North Charleston: Pacific Press, 2012.

Kirsch, Robert W. *The Story of a New Ship of War: L.C.I. (L) ... Landing Craft Infantry (Large) as built at Barber, New Jersey, by New Jersey Ship-building Corporation*. 1997.

Koster, John. *Operation Snow: How a Soviet Mole in FDR's White House Triggered Pearl Harbor*. Washington: Regnery Publishing, Inc., 2012.

Krueger, Walter. *From Down Under to Nippon: The Story of the Sixth Army in World War II*. Combat Forces Press, 1963.

Lukacs, John D. *Escape from Davao: The Forgotten Story of the Most*

Daring Prison Break of the Pacific War. New York: Simon & Schuster, 2010.

MacArthur, Douglas. *Reminiscences*. New York: McGraw-Hill, 1964.

—. *The Campaigns of MacArthur in the Pacific: Volume I (Reports of General MacArthur)*. Washinton D.C.: U.S. Army Center of Military History, 1966.

Manchester, William. *The Glory and the Dream: A Narrative History of the United States, 1932—1972*. Little, Brown, & Company, 1974.

Marquardt, Harold G. *The Following is a Log and Record of My Experiences While in the Service of the United States Navy*. n.d.

McCracken, William. "Can You Top This Sea Story?" *Elsie Item* 80 (2012): 24.

McGee, William L. *The Amphibians Are Coming! Emergence of the 'Gator Navy and its Revolutionary Landing Craft*. Napa: BMC Publications, 2000.

—. *The Solomons Campaigns 1942-1943 From Guadalcanal to Bougainville Pacific War Turning Point*. Napa: BMC Publications, 2001.

Miller Jr., John. *Cartwheel: The Reduction of Rabaul*. Washington: Office of the Chief of Military Department of the Army, 1959.

Morison, Samuel E. *History of United States Naval Operations in World War II*. 15 vols. Little, Brown, & Company, an imprint of Hachette Book Group, Inc., 1966.

Morrill, John H. *The Cincinnati*. Wytheville: Wordsprint, Inc., 1994.

Morris, Frank D. "Bazooka Boats." *Collier's* 11 November 1944.

Morton, Louis. *Strategy and Command: The First Two Years (United States Army in World War II: The War in the Pacific)*. Washington, D.C.: United States Army Center of Military History, 1962.

Naval Institute Press. *Allied Landing Craft of World War Two*. Annapolis: Lionel Leventhal Ltd., 1985.

Nez, Chester and Judith Avila. *Code Talker*. New York: Penguin Group, 2011.

Niedermair, John Charles. *Reminiscences of John C. Niedermair: Oral History* John T. Mason, Jr. U.S. Naval Institute, 1975-1976.

Ohnuki-Tierney, Emiko. *Kamikaze Diaries: Reflections of Japanese Student Soldiers*. Chicago: University of Chicago Press, 2006.

Phillips, Sid. *You'll Be Sor-ree! A Guadalcanal Marine Remembers the Pacific War.* New York: Berkley Caliber/Penguin Group, 2010.

Plant, Louis V. *Memories of World War II.* n.d.

Prange, Gordon, Donald Goldstein and Katherine Dillon. *December 7 1941: The Day the Japanese Attacked Pearl Harbor.* New York: Wings Books, of Random House Co., 1991.

Prefer, Nathan. *Leyte 1944: The Soldiers' Battle.* Havertown: Casemate Publishers, 2012.

Pyle, Ernie and David Nichols. *Ernie's War: The Best of Ernie Pyle's World War II Dispatches.* New York: Touchstone/Simon & Schuster Inc., 1986.

Pyle, Ernie. *Brave Men.* Lincoln: University of Nebraska Press/First Bison Books Printing, 2001.

Reilly Jr., John C. "Lousy Crate Indeed." *Sea Classics* May 1984: 56-85.

Rentz, John N. *Bougainville and the Northern Solomons.* Historical Branch, Headquarters, U.S. Marine Corps, 1946.

Rielly, Robin. *American Amphibious Gunboats in World War II: A History of the LCI and LCS(L) Ships in the Pacific.* Jefferson: McFarland & Company Publishers, 2013.

Robertson, James D. *Robby.* 2012.

Ross, Bill. *Peleliu Tragic Triumph: The Untold Story of the Pacific War's Forgotten Battle.* New York: Random House, 1991.

Rottman, Gordon L. *Landing Craft, Infantry and Fire Support.* Long Island: Osprey Publishing Ltd., 2009.

Shaw, Henry and Douglas T. Kane. *Volume II: Isolation of Rabaul, History of U.S. Marine Corps Operations in World War II.* 1963.

Sherrod, Robert. *Tarawa: The Incredible Story of One of World War II's Bloodiest Battles.* New York: Skyhorse Publishing, 2013.

Sledge, Eugene B. *With the Old Breed at Peleliu and Okinawa.* New York: Ballantine Books, 1981.

Thomas, Kathleen. *Don't Call Me Rosie: The Women who Welded the LSTs and the Men who Sailed on Them.* Tigard: Thomas/Wright, Inc., 2004.

USS LCI Landing Craft Infantry. Vol. 1. Paducah: Turner Publishing Co., 1993.

USS LCI Landing Craft Infantry. Vol. 2. Paducah: Turner Publishing Co., 1995.

Vaughan, Hal. *Sleeping with the Enemy: Coco Chanel's Secret War*. New York: Vintage Books/Random House, 2011.

Weiss, Mitch. *The Heart of Hell: The Untold Story of Courage and Sacrifice in the Shadow of Iwo Jima*. New York: Berkley Caliber, 2016.

Woodward, C. Vann. *The Battle For Leyte Gulf: The Incredible Story of World War II's Largest Naval Battle*. New York: Skyhorse Publishing, Inc., 2007.

Wukovits, John. *For Crew and Country: The Inspirational True Story of Bravery and Sacrifice aboard the USS Samuel B. Roberts*. New York: St. Martin's Press, 2013.

—. *One Square Mile of Hell: The Battle for Tarawa*. New York: NAL Caliber/Penguin Group, 2006.

Zuckoff, Mitchell. *Lost in Shangri-La: A True Story of Survival, Adventure, and the Most Incredible Rescue Mission of World War II*. New York: HarperCollins, 2011.

SOURCES

INTERVIEWS

LCI 22 Crew

Russell P. Sweet, Ensign
- Telephone interview, May 24, 2014

LCI 24 Crew

Louis V. Plant, Signalman
- Telephone interview, December 20, 2013

LCI 64 Crew

James Douglass Robertson, Seaman 2/c (Retired as Major)
- Telephone interview, March 5, 2013
- Author of book, *Robby*

LCI 65 Crew

Stephen Ganzberger, Quartermaster 2/c
- Interview, May 18, 2011
- Home movie videotaped interview, August 3, 1997

J.R. Reid, Seaman 1/c
• Telephone interview, February 5, 2014

LCI 69 Crew
Desmond Howard Johnson, Coxswain
• Telephone interview, January 12, 2021

LCI 70 Crew
Royal Wetzel, Ship's Cook
• Interview with Richard Misenhimer on October 20, 2011 (Transcript provided by National Museum of the Pacific War, Fredericksburg, TX)
• Telephone interview, January 28, 2013
• Telephone interview, February 2, 2013
• Telephone interview, January 24, 2021
John Reulet, Gunner's Mate 3/c
• Interview with Richard Misenhimer on October 4, 2011 (Transcript provided by National Museum of the Pacific War, Fredericksburg, TX)
Leo D. Wilcox, Seaman 1/c
• Interview with Richard Misenhimer on December 1, 2011 (Transcript provided by National Museum of the Pacific War, Fredericksburg, TX)
Gilbert Ortiz, Seaman 1/c
• Email of LCI 70 recollections provided by Joseph Ortiz, Gil's nephew, dated September 15, 2008 at 6:45 p.m.

LCI 74 Crew
Eddie Benoit, Signalman 2/c
• Telephone interview, January 20, 2013

LCI 329 Crew
Elmo Angelo Pucci, Ship's Cook 3/c
• Interview with William L. McGee, date unknown (Transcript

provided by National Museum of the Pacific War, Fredericksburg, TX)

Stephen Ganzberger, Quartermaster 2/c

- Interview, May 18, 2011
- Videotaped interview, August 3, 1997

LCI 331 Crew

Trevor Johns, Petty Officer

- Interview, September 12, 2012

LCI 502 Crew

John P. Cummer, Seaman 1/c (Gunner's Mate)

- Telephone interview, May 27, 2014

LCI 684 Crew

Fred Engelken, Ensign

- Telephone interview, January 14, 2014
- Telephone interview, September 27, 2015

LCI 730 Crew

Dan L. Tolar, Seaman 2/c

- Telephone interview, June 8, 2013
- Telephone interview, July 15, 2013

Army Veterans:

Oliver Jerry Budd, Private 1/c (5th Division of 3rd U.S. Army)

- Interview, August 24, 2014

OFFICIAL REPORTS, WAR DIARIES, DECK LOGS, MUSTER ROLLS

U.S. National Archives Records Administration

USS LCI (L) 65 / USS LCI (G) 65
• Deck Logs, December 16, 1942 to August 15, 1945
• Muster Rolls, January 1944 to May 1945
USS LCI (L) 329
• Deck Logs, March 16 to December 31, 1943
• War Diaries, June to November 1943
• Muster Rolls, March 1943 to January 1944

Rendova, New Georgia (July 1943)
USS LCI (L) 23
• War Diary, July 4, 1943
USS LCI (L) 24
• Action Report, July 4, 1943
• War Diary, July 1943, dated 1 August 1943
USS LCI (L) 63
• War Diary, July 1943
USS LCI (L) 65
• Action Report, anti-aircraft action report of 4 July 1943, dated 11 September 1943
USS LCI (L) 327
• Action Report, anti-aircraft action report, dated 5 July 1943
USS LCI (L) 328
• War Diary, July 1943
USS LCI (L) 329
• Deck Log, month of July 1943
• War Diary, June to July 1943
USS LCI (L) 330
• War Diary, July 1943
USS LCI (L) 332

- War Diary, July 1943
USS LCI (L) 333
- War Diary, July 1943
USS LCI (L) 334
- War Diary, July 1943
USS LCI (L) 335
- War Diary, July 1943
USS LCI (L) 336
- War Diary, July 1943
LCI Group 14
- War Diary, 4 July 1943, dated 19 July 1943
LCI Flotilla Five
- War Diary, June 1943
LST 341
- Action Report, dated 4 July 1943
LST 343
- War Diary, month of July, dated 1 August 1943
LST 396
- Action Report, dated 5 July 1943
LST 397
- War Diary, July 1943
LST 399
- Action Report, dated 5 July 1943
LST 472
- Action Report, for events of 4 July 1943, dated 16 July 1943

Treasury Islands, Bougainville (October/November 1943)
USS LCI (G) 22
- U.S.S. LCI(L) No. 22, Action Report, dated November 5, 1943
USS LCI (G) 23
- Action Report; Support of Landing, Treasury Islands, 27 October 1943
USS LCI (G) 70
- Muster Rolls/Report of Changes, dated 30 November 1943
USS LCI (L) 329

- War Diary, November 1943

Morotai, (September 1944)
USS LCI (G) 65
- Action Report, Morotai Operation, dated 18 September 1944

Leyte, Philippines (October – December 1944)
Commander Task Unit 78
- Action Report, Leyte Operation 1944, Operation Plan No. 101-44
USS LCI (G) 64
- Action Report, Central Philippines – Leyte Operation, dated 1 November 1944
USS LCI (G) 65
- Deck Logs, October to December 1944
- Action Report, Initial Landing Leyte Island, Philippine Islands, dated 13 November 1944
USS LCI (G) 71
- Action Report, Central Philippine Islands, dated 7 November 1944
USS LCI (G) 72
- Action Report, Leyte Operation, dated 1 November 1944
USS LCI (G) 73
- Action Report, Leyte Operation, dated 29 October 1944
USS LCI (G) 74
- Action Report, Central Philippine Islands, dated 3 November 1944
USS LCI (G) 331
- Action report, Central Philippines Operation, dated 5 November 1944
USS LCI (G) 338
- Action Report, Central Philippines Operation, dated 27 October 1944
LCI (L) 432
- Action Report, dated 27 November 1944
USS LCI (G) Group 45
- War Diary, October 1944, dated 1 November 1944
USS LST 66

- Action Report, Leyte Operation, dated 17 November 1944

USS LST 464

- Action Report, Leyte Gulf Operation, 10 October 1944 to 5 December 1944

USS LST 470

- Action Report, Leyte Gulf, dated 15 November 1944

USS LST 552

- Action Report, dated 31 October 1944

USS LST 555

- War Diary, month of October 1944, dated 13 November 1944

USS *Achilles* (ARL-41)

- Action Report – Anti-Aircraft, Leyte Operation, narrative form, dated 12 November to 16 November 1944

USS *Blue Ridge* (AGC-2)

- Action Report, Leyte Operation, dated 15 November 1944

USS *Quapaw* (ATF-110)

- Action Reports – Philippines Operation, dated 29 October to 14 November 1944

Lingayen Gulf, Luzon, Philippines (January 1945)

USS LCI (G) 64

- Action Report, Luzon Operation – 5 January 1945, dated 15 January 1945
- Action Report, Luzon Operation – 9 January 1945, dated 15 January 1945

USS LCI (G) 65

- Action Report, Lingayen Operation, 2 to 7 January 1945
- Action Report, Lingayen Operation, 9 January 1945

USS LCI (G) 68

- Action Report, Luzon Operation – 1 to 6 January 1945, dated 26 January 1945
- Action Report, Luzon Operation – 9 to 13 January 1945, dated 26 January 1945

USS LCI (G) 69

- Action Report, Luzon, 5 January 1945

- Action Report, Luzon, 6 January 1945
- Action Report, Lingayen Operation (minesweeping phase), dated 10 February 1945

USS LCI (G) 70

- Action Report, Lingayen Operation, dated 17 January 1945
- Muster Rolls/Report of Changes, dated 30 December 1944 and 31 January 1945

USS *Apache* (ATF-67)

- Action Report of 5 January 1945
- Action Report of 6 January 1945

PERSONAL ACCOUNTS

Benoit, Eddie, various deck logs, personal accounts
• Undated documents include: personal accounts, deck logs, a printout of the LCI 74's navsource.org web page, and a copy of "LCI Blues" poem. Included a DVD of his interview with Veteran's History Project dated 2007.

Block, Leonard, Jr., "USN WWII – LCI971 – Memoirs"
• Undated four-page typewritten personal account based on experiences aboard LCI 971.

Cummer, John P., "USS LCI 502 at D-Day"
• Undated seven-page typewritten personal account based on personal recollections. Included was the deck log of USS LCI 502.

Harris, Joe L., "My Adventures In WWII"
• Undated typewritten ten-page personal memoirs of his experience aboard LCI 600 (September 15, 1943–April 26, 1946). Also included photos and newspaper clippings.

Hartwell, Russell W., "Amphibious Assault Landing Craft U.S.S. LCI(L) 711"
• One hundred sixty-page typewritten personal memoirs of his experiences as a signalman aboard the LCI 711 (Flotilla 24; 7th Amphibious Fleet; Philippines; World War II). Castle Publishers.

Johns, Trevor R., unnamed personal account
• Twenty-three-page typewritten story of his WWII experience aboard LCI (R) 331, dated May 1, 1990. Also included LCI 331's three-page official Action Report from Balikpapan Operation on July 1, 1945, muster roll from LCI 331 dated June 30, 1945, and various other timelines of LCI 331.

Keeler, William E., "Report of LCI Duty from 1944 to 1947"
• Undated twenty-page typewritten reminiscences of life as an Ensign aboard LCI 448. Retired from U.S. Navy as a Commander.

Kirsch, Robert W., booklet, "The Story of a New Ship of War: L.C.I. (L) ... Landing Craft Infantry (Large) as built at Barber, New Jersey, by New Jersey Shipbuilding Corporation"
• Booklet edited and published by Robert Kirsch, who served aboard LCI (R) 74. This booklet was given to each LCI crewman who attended the USS LCI National Association's 7th reunion at the Hilton Hotel in Cherry Hill, N.J., held April 23–27, 1997.

Marquardt, Harold G., "The Following is a Log and Record of My Experiences While in the Service of the United States Navy"
• Twenty-nine-page personal record and daily log of his experience aboard the LCI 15. Provided by John P. Cummer, LCI 502.

Ortiz, Gilbert, personal recollections in email
• Email to nephew, Joseph Ortiz, dated September 15, 2008 at 6:45 a.m. (Subject: Re: Joey from Texas). Included LCI 70 memories of kamikaze attack in January 1945 at Luzon.

Plant, Louis V., Personal memoirs, "Memories of World War II"
• Undated fifty-five-page typewritten story of his WWII experience in the Amphibious Force, including life aboard LCI 24.

Tolar, Dan L., generic documents
• Undated seven-page typewritten reminiscences of life under Commodore John H. Morrill as a seaman aboard LCI 730. See also: Dan's stories in Morrill's book, *The Cincinnati*.

Wollard, James, "History of USS LCI(G)345 – An Amphibious Ship Serving in the Pacific Theater in World War II" (Copy No. 10), Gunner's Mate 1/c
• A collection of typewritten accounts from various sailors who

served aboard LCI 345 in World War II, and included photos and official action reports. LCI 345 veterans in this collection of personal accounts included: Morton Myerson, Charles P. Grow, Elmer Dire, Ray Paradise, Jim Wollard, Don Reinke, Arthur H. Lillibridge, and James G. Atteberry. Provided by John P. Cummer, LCI 502.

ELSIE ITEM ISSUES

(*Note: Elsie Item* is the official newsletter of the USS LCI National Association)

Issue 11 (December 1994)
• Frank D. Morris, "Bazooka Boats," pp. 7–8. Originally from *Collier's* (November 1944 issue).
Issue 12 (March 1995)
• Winston LeRoy, "An LCI Creed," p. 4.
• Charles H. Uhl, Executive Officer, "Boston to Borneo—LCI (M)362," p. 7.
Issue 14 (September 1995)
• Dr. R. William Clark, "LCI's with Sails?" p. 12.
Issue 15 (December 1995)
• Thomas E. Woodstrup, "LCI (FF) 628," p. 13–14. Originally published in *American Legion* (September 1991 issue).
Issue 17 (June 1996)
• Robert V. Rosenwald, "USS LCI(L) 1008," pp. 13–14.
Issue 21 (June 1997)
• W. Donald Stewart, "LCI(R) 706," p. 12.
Issue 26 (September 1998).
• Sgt. Bill Alcine, "A Lucky Japanese Dive-Bomber Spoils the Day for the LCI (G) #23," pp. 27–28. Originally published in *Yank*, Down Under Edition (1944 issue).
Issue 27 (December 1998)
• Howard G. Sawyer, "This is the Story of the LCI 336," p. 23–24.
Issue 32 (March 2000)
• Garnett A. Ridenhour (LCI 638), "The War Years," pp. 21–23. Submitted by Uma Ridenhour in memory of her husband, Garnett, who passed away February 14, 1998.
• Harold R. Ronson (LCI 1012), "Remembering Franklin Delano Roosevelt," p. 34.
Issue 41 (June 2002)

• John P. Cummer, "Who We Were . . . Who We Are: Some Memorial Day Thoughts from the Editor," p. 4.

Issue 43 (January 2003)

• "Big Bill" Athan, "The Saga of the LCI 446: Salty Tales of the South Pacific!" p. 23.

Issue 50 (October 2004)

• Chuck Savard, "An LCI at the Battle of Leyte Gulf – and other places!" pp. 29–30.

Issue 51 (December 2004)

• John Cummer with Chris Shelvik (LCI 337) and Ardie Hunt (LCI 226), "LCIs and the Rescue Operations at the Battle of Samar 'The Rest of the Story,'" pp. 18–22.

Issue 53 (April 2005)

• Elmer Kinsinger, "Letters to the Editor," p. 6.

Issue 56 (June 2006)

• Charles R. Ports, "My Life as an LCI Gunboat Sailor," pp. 19–22.

Issue 61 (October 2007)

• Patricia Rone, "Letters to the Editor," p. 9.

Issue 64 (August 2008).

• Bob L. Petit, "Bob Petit's Story: The Trials and Triumphs of an Underage LCI Sailor!" p. 18–21.

Issue 71 (May 2010)

• Dominick Maurone, "Voices from the Pacific," pp. 23–27.

Issue 72 (July 2010)

• George Weber (LCI 370), "Life Aboard a Flotilla Flagship or It ain't quite exactly what you thought!" p.11.

Issue 75 (May 2011)

• Albert D. Divincenzo, "My Wartime Experiences Aboard USS LCI (G) 373," pp. 20–23.

Issue 77 (October 2011)

• Jim Finkler/William D. Elder, p. 9

Issue 79 (June 2012)

• Robert Harker, "The LCI Gunboats at Iwo Jima," p. 17.

Issue 80 (December 2012)

• William H. McCracken, "Can You Top This Sea Story?" p. 24.

Issue 81 (February 2013)

• Richard W. Johnston (United Press), "Gallantry at Eniwetok," pp. 23–24.

Issue 86 (May 2014)

• Zach Morris, "On the Fourth They Fought Back," pp. 9–15.

Issue 89 (February 2015)

• Zach Morris, "The LCI Gunboats of Iwo Jima D–2: February 17, 1945," pp. 12–20.

Issue 90 (May 2015)

• "We Remember President Franklin Roosevelt," pp. 10–11.

Issue 92 (November 2015)

• Zach Morris, "The LCIs in the Battle for the Philippines," pp. 6-10; "He Was Headed Right For Me," pp. 15–19.

Issue 93 (April 2016)

• "Vice Admiral Lorenzo Sherwood Sabin Jr. May 23, 1899–June 2, 1988," pp. 12–13.

WEBSITES

Arlington National Cemetery – Stephen Ganzberger's Gravesite:
• https://ancexplorer.army.mil/publicwmv/#/arlington-national/
search/results/1/CgpHYW56YmVyZ2VyEgdTdGVwaGVu/

Barger, Melvin D., "The Genius Behind The LST: John C. Niedermair Was a Giant in the Field of Naval Architecture":
• https://www.uslst.org/memories/27-articles/16-lst-memories-the-genius-behind-the-lst

D-Maps:
Pacific Ocean:
• https://d-maps.com/carte.php?num_car=3260&lang=en
Papua New Guinea
• https://d-maps.com/carte.php?num_car=3336&lang=en
Philippines:
• https://d-maps.com/carte.php?num_car=590&lang=en
Terms & conditions for commercial use:
• https://d-maps.com/conditions.php?lang=en

Dingell, John D., House page photograph:
• https://history.house.gov/Collection/Detail/15032404848

General Quarters and condition type details:
• https://seabeemagazine.navylive.dodlive.mil/files/2014/02/Chapter-3.pdf

Green Islands Info:
• https://www.history.navy.mil/content/history/nhhc/research/library/online-reading-room/title-list-alphabetically/b/building-the-navys-bases/building-the-navys-bases-vol-2.html#1-25

Hornfischer, James D., *Smithsonian Magazine*, "Revisiting Samuel Eliot Morison's Landmark History" (February 2011):
• https://www.smithsonianmag.com/history/revisiting-samuel-eliot-morisons-landmark-history-63715/

LCI Facts:
• http://usslci.org/facts/

LCI 65 Info:
• https://www.navsource.org/archives/10/15/150065.htm

LCI 329 Info:
• https://www.navsource.org/archives/10/15/150329.htm

Lingayen Gulf, U.S. Army Map (public domain):
• https://ww2db.com/image.php?image_id=24858

Navsource (LCI list):
• http://www.navsource.org/archives/10/15/15idx.htm

New Georgia Map (public domain):
• https://www.ibiblio.org/hyperwar/USMC/II/maps/USMC-II-II.jpg

Nichols, Robert, "The First Kamikaze Attack?" (May 20, 2020):
• https://www.awm.gov.au/wartime/28/kamikaze-attack

Notre Dame Alumnus, The (October 1936 issue), Archives of the University of Notre Dame:
• http://www.archives.nd.edu/Alumnus/VOL_0015/VOL_0015_ISSUE_0001.pdf

Philippines Information:
• https://www.nationalgeographic.com/travel/article/partner-content-know-before-you-go-the-philippines

DVDs

Home movie, Ganzberger Family Collection, August 3, 1997.

Veteran's History Project – Ed Benoit (LCI 74), Interview & Montage, dated 2007. Produced by Dixon Schwabl.

ABOUT THE AUTHOR

Zach Morris is currently the editor-in-chief of *LST Scuttlebutt*, the quarterly magazine newsletter of the U.S. Landing Ship Tank (LST) Association. He was also the editor-in-chief of *Elsie Item*, the quarterly magazine newsletter of the USS Landing Craft Infantry (LCI) National Association from 2014–2016. Morris, a graduate of Michigan State University, is an award-winning senior financial analyst and certified screenwriter in Los Angeles. This is his first book.

instagram.com/zach_editor

tiktok.com/@zach_editor

youtube.com/@zach_editor